Generis
PUBLISHING

ORGANIC PRODUCTION IN THE REPUBLIC OF SERBIA

Dusan Dabovic

Title: **ORGANIC PRODUCTION IN THE REPUBLIC OF SERBIA**

ISBN: 979-8-89966-137-2

Author: Dusan Dabovic

Cover image: www.pixabay.com

Publisher: Generis Publishing
Online orders: www.generis-publishing.com
Contact email: info@generis-publishing.com

CONTENTS

Introduction

According to the definition, organic production encompasses a set of agricultural methods for food production that utilise materials and processes that contribute to the protection of biodiversity and minimise pollution of water, air, and soil.

Organic agriculture in Serbia began in the 1990s, when the first crops were certified as organic and when the association “Tera” was founded in Subotica. This was the first association in Serbia to focus on organic production, launching and initiating numerous activities to promote and develop organic agriculture. In 2005, “Tera” organised the first festival of organic products, “Biofest”, which has since become a traditional international event held in autumn. This first festival gathered over 80 exhibitors from more than 20 cities in Serbia and abroad (Croatia, Hungary, Bosnia and Herzegovina, Germany, and Italy), and around 500 participants took part in various accompanying programs at this event.

In the Republic of Serbia, organic production was first regulated at the federal level by the Law on Organic Agriculture, enacted in 2000. Then followed the law of the Republic of Serbia from 2006, and now the valid law from 2010, with the latest amendments from 2017, is in force. This law regulates the production of agricultural and other products using organic production methods, the goals and principles of organic production, organic production methods, control and certification in organic production, processing, labelling, storage, transportation, traffic, import and export of organic products, as well as other issues of importance for organic production.

Based on this law, two regulations were adopted, which regulate professional individual problems in more detail. Thus, the Rulebook on control and certification in organic production and methods of organic production prescribes in more detail the conditions that must be met by the control organisation for the performance of control and certification in organic production, methods of organic crop and livestock production, the duration of the conversion period in crop and livestock production, the method of control in organic production and the measures that the authorised control organisation determines in the event of established irregularities in the method of carrying out the production of organic products, the content and form of the records kept by the authorised control organisation, as well as the way of keeping those records, the content and form of collective records, technological procedures of processing in organic production, ingredients, additives and auxiliary substances used in the processing of organic products, the method of cleaning and means for cleaning technological lines, the appearance of the mark and the national sign of organic products, as well as other specific issues in this area.

Also, the Rulebook on the documentation to be submitted to the authorised control organisation for the purpose of issuing a certificate, as well as the conditions and method of selling organic products, regulates the documentation to be submitted to the authorised control organisation for the purpose of issuing a certificate that the certified organic product that is imported was produced in accordance with the law regulating organic production and the regulations adopted on the basis of that law, as well as the conditions and method of selling organic products that can be sold together with products from conventional production only if they are packaged.

Additionally, in terms of development, organic production in the Republic of Serbia was regulated on a strategic basis for the period 2014–2024. This document states that our country offers favourable conditions for the development of organic production, considering its natural resources, favourable land and climate conditions, biodiversity, and relatively healthy agroecosystems. However, it should be noted that (at the time of writing this act), an appropriate legal framework defining organic production has not been developed, although its great importance and potential are recognised.

At the same time, it is stated that the domestic market of organic products is not sufficiently developed since, until then, sufficient efforts were not undertaken to educate producers and consumers and promote organic products, and only organic products of plant origin are exported from the Republic of Serbia, and mainly partially processed, so that the most represented export at that time was frozen fruit (72% of the total amount of organic products).

Based on this long-term strategy, medium-term strategic documents were adopted that included the development of organic production. Given the emphasis on the previous long-term strategic framework, it can be expected that in 2025, a new long-term strategy for agriculture will be adopted, along with medium-term development programs, including organic production.

In this study, after analysing the regulations that make up the legal framework on organic production in the Republic of Serbia, we presented a corresponding report on the application of regulations on organic production. Thus, a corresponding report for the period up to 2020 was presented. This report presents an overview and development of organic production in Serbia, legislative and institutional framework, view of the domestic market of organic products, agricultural policy and rural development policy on the way to the EU, SWOT analysis, development of the private sector in less developed regions of Serbia, as well as other issues in this area. At the end of the text we presented analysis of statistical data of organic production in the Republic of Serbia.

I. Law on organic production in the Republic of Serbia

1. Basic rules on organic production

a) General provisions

In the Republic of Serbia, organic production of agricultural and other products is regulated by a special law, and within its framework, the objectives and principles of organic production, organic production methods, control and certification in organic production, processing, labeling, storage, transportation, trade, import and export of organic products, as well as other issues of importance for organic production.[1] In addition, it has been established that the provisions of this law apply to products originating from all stages of organic plant and livestock production, including animal aquaculture, and when these products are intended for trade, and refer to: primary agricultural products; processed agricultural products used as food; animal feed; seeds, seedlings and planting material; yeasts used as food and animal feed. Therefore, products obtained from hunting wild animals and fishing are not considered products of organic production.[2]

As stated, the objectives of this law are: establishing organic production as a comprehensive system of management and food production based on ecological practices, a high level of biological diversity (biodiversity), the conservation of natural resources, the application of high standards of animal welfare and production methods that include the use of natural substances and processes; balanced plant and livestock production that respects natural systems and cycles, maintains and improves soil fertility and quality, water and air quality; rational use of energy and natural resources, such as soil, water, organic matter, etc.; production of various organic agricultural products in accordance with consumer demands for organic food, using processes that are not harmful to human, plant, animal health and the environment as a whole. In doing so, it is stated that the basic principles in organic production are: a) the principle of management of natural and biological processes, which refers to the management of natural and biological processes based on ecological systems of natural resource use, with the application of methods that include: living organisms and mechanical production methods, plant production in soil, livestock production or aquaculture that

---

[1] The Law on Organic Production (Official Journal of the Republic of Serbia, No. 30/10, 17/19 - other law (in Serbian)).
[2] *Ibid.*, Art. 1, 2.

respects the principles of sustainable exploitation of fish farms, the prohibition of the use of genetically modified organisms and products consisting of or derived from genetically modified organisms (with the exception of veterinary medicinal products), production processes that are based on risk assessment and appropriate preventive measures (where necessary); b) the principle of limited use of input materials (external inputs), which refers to the limitation of the use of input materials in organic production, whereby inputs from organic production, natural or naturally produced substances, and slow-release mineral fertilizers may be used; c) the principle of strict limitation of the use of chemical synthesized inputs in organic agriculture, except in cases where appropriate management systems do not provide satisfactory results, or if the aforementioned external inputs are not available on the market, or if the use of the aforementioned external inputs has an unacceptable impact on the environment; d) the principle of adaptation, which refers to the adaptation of organic production methods to regional and local climatic and agro-ecological conditions, sanitary conditions, level of development and specificities of traditional farming methods; d) the principle of maintaining and improving life in the soil and natural soil fertility, soil stability and biodiversity that prevents and disables soil compaction and erosion, as well as plant nutrition primarily through the soil ecosystem.[3]

b) Entities performing tasks related to organic production

As stated in this law, in order to consider expert issues, provide expert opinions and participate in the implementation of project tasks in the field of organic production, the minister responsible for agricultural affairs, in accordance with the regulations governing state administration, shall establish a special working group by decision - the Expert Council for Organic Production. The Expert Council shall consider expert issues, provide expert opinions and participate in the implementation of project tasks related to: plans and special programs in the field of organic production; by providing recommendations for professional training of personnel; by providing recommendations for the development of regulations in the field of organic production; by providing recommendations for the improvement of organic production, and by performing other necessary tasks related to organic production.[4]

[3] *Ibid.*, Art. 5, 6.

[4] *Ibid.*, Art. 7.

c) Control organizations

As prescribed by this Law, the Ministry may authorize a control organization registered in the Register of Business Entities in accordance with the law governing the registration of business entities, which has an accreditation act issued by the competent accreditation body of the Republic of Serbia in accordance with the relevant Serbian, European and international standards, and which meets the requirements in terms of technical equipment and personnel training, to perform control and certification tasks in organic production. In doing so, the Minister shall determine by decision whether the conditions for performing the aforementioned control and certification tasks in organic production are met, as well as the code number of the authorized control organization. This decision of the Minister shall be issued for a period of one year and may be extended in accordance with the conditions provided for in this Law. The decision shall be final and an administrative dispute may be initiated against it. In addition, the Minister is authorized to prescribe in more detail the conditions that must be met by the control organization for performing control and certification tasks in organic production.[5]

As stated, the control organization submits to the Ministry a request for determining the fulfillment of the conditions for performing control and certification tasks in organic production by 30 June of the current year for the following year. The authorized control organization may, within the given deadline, submit a request for renewal of the authorization for performing control and certification tasks in organic production. In that case, the Ministry determines whether the authorized control organization still meets the conditions for performing control and certification tasks in organic production. Then, the Ministry determines the list of authorized control organizations and publishes it once a year in the "Official Gazette of the Republic of Serbia". In addition, this ministry keeps records of authorized control organizations, which contain data on the name and address of the authorized control organization; the date of submission of the application for granting or renewing authorization; the date of adoption of the decision on authorization or renewal of authorization; the code number of the authorized control organization; any changes in relation to the fulfillment of the conditions for performing control and certification tasks in the field of control and certification in organic production. These records are kept in electronic form. The law stipulates that the authorized control organization is obliged to notify the ministry in writing without delay of any change in the conditions for performing control and certification tasks in organic production and to submit evidence of the

---

[5] *Ibid.*, Art. 9.

change. If an authorized control organization ceases to meet the conditions for performing control and certification tasks in organic production, as well as at the request of the authorized control organization, the Minister shall, by decision, revoke the authorization of the control organization to perform control and certification tasks in organic production. This decision of the Minister is final and an administrative dispute may be initiated against it.[6]

d) Organic production methods

According to the law, organic plant production methods include the selection of plant species and varieties; crop rotation; soil cultivation systems; means and methods of fertilization; soil fertility maintenance system; methods of controlling plant diseases, pests and weeds; methods of collecting wild plant species from natural habitats. In addition, the Minister is authorized to prescribe in more detail the methods of organic plant production. It is prescribed that organic livestock production methods include the selection of animal species and breeds; methods of raising animals and facilities for raising animals; nutrition and health care of animals; transport and slaughter of animals; treatment of animals acquired from other farms; methods of collecting animal species from natural habitats. The Minister is also authorized to prescribe in more detail the methods of organic livestock production. It is stated that reproductive material for plant and livestock production used in organic production must be produced using organic production methods. Therefore, seed production using organic production methods is carried out in accordance with the regulations governing the production of seeds, seedlings and planting material, using organic production methods. In special cases, the producer may be approved to use reproductive material from conventional production, provided that this material has not been treated with plant protection products that are not permitted in organic production. This exception applies to the following cases: if there is no material produced using organic production methods on the market; if the use of this material is justified for conducting scientific research; and if it is an indigenous variety. The use of reproductive material from conventional production during the conversion period is approved by the producer by the authorized control organization, and after the conversion period has expired, by the relevant ministry.[7]

[6] *Ibid.*, Art. 10-13.

[7] *Ibid.*, Art. 14-16.

e) Control and certification in organic production

It is envisaged that the procedure for including a producer in organic production is initiated by an application submitted by the producer to the authorized control organization. Based on this application, the producer and the authorized control organization conclude a contract on the performance of control and certification in organic production, which determines the registration number that the producer receives. The conversion period begins on the date of conclusion of this contract, and the Minister shall prescribe in more detail the duration of the conversion period in plant and livestock production. The producer who has concluded the aforementioned contract is obliged to carry out organic production in accordance with this Law and the regulations adopted on the basis thereof from the date of conclusion of that contract. In doing so, the producer is obliged to keep records of his production in accordance with the instructions of the authorized control organization, as well as to inform the authorized control organization in writing every year, within the deadline set by it, about the dynamics of crop production, along with submitting a survey by parcel. In any case, the producer is obliged to enable the authorized control organization to carry out control. If the producer also has a production unit used for conventional production within the agricultural holding, he is obliged to enable the authorized control organization to carry out control of that production unit as well.[8]

It is stipulated that control tasks in an authorized control organization may be performed only by a person who is professionally qualified to perform control in organic production. In addition, the aforementioned control tasks may not be performed by a controller if he is also a producer involved in organic production or is engaged in the trade of organic products, or if he performs control tasks in more than one control organization. The authorized control organization is obliged to take the necessary measures to ensure the confidentiality of information obtained during the implementation of control and certification. Control in organic production is carried out in a manner that ensures objectivity, impartiality, absence of conflict of interest, efficiency, professionalism, consistency and quality in all phases of organic production and for all given products, by applying an appropriate control method and based on a documented procedure. If during the aforementioned control it is observed that there are certain irregularities in the method of carrying out the production of organic products, the authorized control organization may determine corrective measures for the producer in order to harmonize production with the provisions of this Law and the regulations adopted on the basis thereof, whereby the Minister shall prescribe in more

[8] *Ibid.*, Art. 17, 18.

detail the method of carrying out control in organic production, as well as the measures that the authorized control organization determines in the event of established irregularities in the method of carrying out the production of organic products. The authorized control organization is obliged to carry out a full physical control and control of the producer's documentation at least once a year, and to draw up a written report on the control carried out, which is also signed by the producer.[9]

Based on the report on the controls carried out, the authorized control organization issues a certificate that the product or production process complies with this Law and the regulations adopted on its basis. For certified organic products that are imported, the authorized control organization, based on the performed control of the documentation and the certificate issued by the competent authority of the country of origin of the certified organic product, issues a certificate that the product was produced in accordance with this Law and the regulations adopted on its basis. In the procedure for issuing the said certificate, the importer is obliged to submit to the authorized control organization the necessary documentation on the basis of which it can be established that the product was manufactured in accordance with this Law and the regulations adopted on the basis thereof, and that a certificate has been issued for it by the competent authority of the country of origin. The authorized control organization performs these tasks as entrusted tasks. In the case of issues not specifically regulated by this Law, and relating to the procedure for issuing certificates and certificates, the provisions of the law governing general administrative procedure shall apply. In this regard, the Minister shall prescribe in more detail the documentation to be submitted for the purpose of issuing the said certificate.[10]

The law stipulates that the authorized control organization is obliged to keep records that contain, in particular, data on: the name and address, or business name and registered office of the producer, as well as data on the person responsible for organic production; the address of the organic production unit; the date of the start of the conversion period; the code number of all control organizations with which the organic producer was registered after inclusion in organic production; arable areas under organic production; the area from which wild plant and animal species were collected from natural habitats, in accordance with the regulation governing the collection and use of wild plants and animals; the type and number of animals on the agricultural holding; issued certificates and receipts; production of reproductive material. In addition, the control organization is obliged to keep these records for at least three

[9] *Ibid.*, Art. 19–21.
[10] *Ibid.*, Art. 22.

years, as well as to provide data from these records at the request of the ministry. It is stated that the Minister shall prescribe in more detail the content and form of the said records, as well as the manner of keeping those records. In addition, the authorized control organization shall be obliged to submit to the Ministry every year, no later than 31 January of the current year, an annual report for the previous year on organic production, which shall contain data from the said records. Furthermore, based on the data from the said annual reports, the Ministry shall keep summary records on organic production, which shall be kept in electronic form, while the Minister shall prescribe in more detail the content and form of the said records.[11]

f) Processing, labeling, storage, transportation, trade, import and export of organic products

Processing in organic production takes place continuously without mixing products from organic production with ingredients, substances and products from the conversion period, i.e. products from conventional production, as well as without the use of ionizing radiation, genetically modified organisms and their derivatives. If the processing process uses technological lines that are also used for processing products from conventional production, these technological lines must be thoroughly cleaned and washed before use in organic production. The Minister shall prescribe in more detail the technological processes of processing in organic production, the ingredients, additives and auxiliary substances used in the processing of organic products, as well as the method of cleaning and cleaning agents for technological lines outside the processing of organic products.[12] Regarding the labeling of organic products, the following is prescribed: a) a domestic unprocessed certified organic product shall be labeled with the label "ORGANIC PRODUCT", the national mark and the code/logo of the authorized control organization that certified that product; b) products of plant origin from the conversion period that contain exclusively one ingredient of agricultural origin may, after the expiration of the deadline, i.e. one year from the date of conclusion of the organic production contract, be labeled with the label "PRODUCT FROM THE CONVERSION PERIOD" and the code/logo of the authorized control organization; c) a domestically processed certified organic product may be marked with a label for an unprocessed product, a national mark and the code/logo of the authorized control organization that certified the product in the last stage of production,

[11] *Ibid.*, чл. 23–25.

[12] *Ibid.*, чл. 26.

if it contains at least 95 percent of ingredients of agricultural origin that were produced in accordance with the provisions of this Law and the regulations adopted on the basis thereof. In addition, advertising materials, accompanying documentation and labels may also be marked in the manner referred to in points a) and b). Products that indicate that they contain genetically modified organisms, consist of or are produced from genetically modified organisms may not be marked in the manner referred to in points a) and b). Certified products from points a) and c), as well as their advertising material, accompanying documentation and labels may be marked with terms or abbreviations indicating the origin of the organic product (ecological, biological, organic, or eco, bio, org., etc.), as well as with the private label of the producer or producer association. It is prescribed that the Minister prescribes the appearance of the label and the national logo for organic products.[13]

Certified organic products must be stored in special warehouse premises. If certified organic products are packaged and labeled, they may be stored in the same warehouse premises with products from conventional production, but in a separate labeled section. The Minister shall prescribe in more detail the method of storing organic products. Also, as with storage, certified organic products shall be transported separately from products from conventional production. In addition, if certified organic products are packaged and labeled, they can be transported together with products from conventional production. The Minister shall also prescribe in more detail the method of transportation of organic products.[14]

Regarding the trade in organic products, it is stipulated that organic products may be sold together with conventionally produced products only if they are packaged. In addition, if organic products are not packaged, they must be sold separately from conventionally produced products and in a clearly marked place, in order to prevent mixing with conventionally produced products. In any case, the producer is obliged to have a valid certificate issued by an authorized control organization when placing organic products on the market. By-laws on the conditions and manner of sale of organic products are issued by the minister of the competent authority. Regarding the import of organic products into our country, it is stipulated that certified organic products must be packaged and closed in a way that prevents substitution of their contents. These products may be placed on the market if the authorized control organization of the given country has issued the appropriate certificate for these products. As with the trade in organic products, the Minister of the competent authority

---

[13] *Ibid.*, Art. 27.

[14] *Ibid.*, Art. 28, 29.

shall issue a by-law that further regulates the conditions for the import of organic products. In addition, it is stipulated that certified organic products may be exported only with a valid certificate from an authorized control organization. Namely, the exporter shall enter in the appropriate section of the customs declaration that it is an organic product and the number of the certificate attached to that customs declaration.[15]

g) Supervision

Inspection supervision over the implementation of the provisions of this Law and the regulations adopted on its basis shall be carried out by the competent ministry through inspectors responsible for the supervision of organic production. Also, supervision over the work of authorized control organizations in the performance of entrusted tasks shall be carried out by the competent ministry in accordance with the regulations governing state administration. In the performance of inspection supervision tasks, the inspector for organic production shall have the rights and duties prescribed by this Law.[16]

[15] *Ibid.*, Art. 30–32.

[16] *Ibid.*, чл. 33–36.

II. Rules on control, certification and methods in organic production

1. Introductory provisions

In the Republic of Serbia, a special regulation regulates in more detail the conditions that must be met by a control organization for carrying out control and certification tasks in organic production, the methods of organic plant and livestock production, the duration of the conversion period in plant and livestock production, the method of performing control in organic production and the measures that the authorized control organization determines in the event of established irregularities in the method of carrying out the production of organic products, the contents.[17] This regulation also regulates the form of records kept by the authorized control organization, as well as the method of keeping those records, the content and form of summary records, technological processing procedures in organic production, ingredients, additives and auxiliary substances used in the processing of organic products, the method of cleaning and cleaning agents for technological lines, the appearance of the label and national symbol of organic products, as well as the method of storage and transportation organic products.[18]

2. Conditions to be met by a control organization for performing control and certification tasks in organic production

A control organization registered in the Register of Business Entities, in accordance with the law regulating organic production, may be authorized to perform control and certification tasks in organic production if it is registered in the Register of Business Entities, in accordance with the law regulating the registration of business entities, and which meets the following conditions in terms of accreditation, technical equipment and personnel training: it is accredited by the competent accreditation body of the Republic of Serbia in accordance with the SRPS EN ISO/IEC 17065 standard; it has an established organizational structure, with a description of the tasks performed, as well as an established method of performing control and certification tasks in organic production; it has an established standard control procedure containing a

[17] Regulation on control and certification in organic production and organic production methods (*Official Journal of the Republic of Serbia*, No. 95/20, 24/21(in Serbian)).
[18] *Ibid.*

detailed description of control measures, precautionary measures and corrective measures taken during control; has office and telecommunications equipment with an internet connection, necessary for carrying out control and certification tasks in organic production; has at least two persons employed to carry out control and certification tasks in organic production, one of whom is in control tasks and the other in certification tasks. The person in certification tasks has higher education acquired through second-level studies (master of academic, specialist academic or master of professional studies) or basic studies lasting at least four years, in the fields of agriculture, technological engineering (food technology and biochemistry) or veterinary medicine. He may also hold a doctorate in biotechnical, technological, medical (veterinary medicine), biological, environmental or biochemical sciences. In addition, he must have appropriate experience in control or certification tasks in organic production and completed training in the field of organic production.

According to the prescribed procedure, the control organization submits to the Ministry a request for determining whether the conditions for performing control and certification tasks in organic production are met, as well as a request for renewal of the authorization for performing control and certification tasks in organic production, by 30 June of the current year for the following year, in accordance with the law regulating organic production. The request for determining whether the conditions for performing control and certification tasks in organic production are met shall be accompanied by: proof of registration in the Register of Economic Entities; accreditation act issued by the competent accreditation body of the Republic of Serbia; act on the organization and description and method of performing control and certification tasks in organic production; proof of the established standard control procedure containing a detailed description of control measures, precautionary measures and corrective measures taken during control; proof of possession of office equipment and telecommunications equipment with an internet connection, necessary for performing control and certification tasks in organic production; employee data (name and surname, professional qualifications, work experience, professional development, or training in the field of organic production, procedure and method of selecting personnel, procedure for appointing and working of expert committees); internal act on data protection obtained in the process of performing control and certification tasks; contracts concluded by the control organization with third parties in connection with performing control and certification tasks in organic production; forms of documents issued by the control organization in the control and certification procedure; statement of the responsible person on the absence of a conflict of interest in connection with performing control and certification tasks in organic production; other documentation if necessary. The request for renewal of authorization to perform control and certification tasks in organic production shall be accompanied by: an annual report on the performed professional control and certification and decisions on complaints and disputes from the previous year; a report on changes in the organization of the authorized control organization during the previous year; other documentation if necessary. The ministry obtains certain documentation ex officio in accordance with the law regulating general administrative procedure.[19]

3. Organic production methods

According to this regulation, it is stipulated that organic production may not use food, animal feed, processing aids, plant protection and nutrition products, soil improvers, reproductive material, microorganisms or animals that are GMOs or products obtained from or by GMOs. The origin of a product used as food or animal feed shall be determined on the basis of the label on the product or on the basis of an accompanying document attached or delivered with the product, and in accordance with the law governing food safety, or the regulation governing the declaration, labelling and advertising of food. Products that are not used as food or animal feed and originate from conventional production, but are used in organic production, must be accompanied by a statement from the seller confirming that the delivered products are not produced from GMOs or by GMOs. In addition, it is stipulated that ionizing radiation may not be used in the processing of organic food or animal feed, or raw materials used in the production of organic food or animal feed.

4. Application of organic production methods

Organic production methods include organic plant production methods and organic livestock production methods. The application of the above methods ensures: maintenance and improvement of the soil, higher organic matter content, physical, chemical and microbiological activity of the soil, its stability and biological diversity that prevents soil compaction and erosion, as well as plant nutrition through the soil ecosystem; minimal use of non-renewable resources and raw materials that do not originate from the same agricultural property; recycling of waste and by-products of plant and animal origin as raw materials in plant and livestock production; consideration of local or regional agro-ecological conditions when making decisions

[19] *Ibid.*, Art. 4.

on the type and method of production; maintaining animal health by stimulating natural immunity, as well as by selecting appropriate species, breeds and breeding practices; maintaining plant health by applying preventive measures, such as selecting appropriate species and varieties resistant to pests and diseases, appropriate crop rotation, mechanical and physical plant protection practices, as well as applying biological control measures for pathogens, pests and weeds, which also includes protecting the natural enemies of pests of cultivated species; animal breeding practices that are adapted to the local conditions in which production takes place and sufficient open areas used by animals; respecting a high level of animal welfare, taking into account the specific needs of each species, i.e. reducing any suffering and mutilation of animals throughout their lives, including at the time of slaughter, to the minimum possible extent; production of organic livestock products by using animals that have spent their entire lives on farms engaged in organic production, from birth or hatching; selection of species and breeds, taking into account the ability of animals to adapt to local conditions, their vitality and resistance to diseases or health problems; feeding animals with feed consisting of agricultural ingredients of organic origin and natural non-agricultural materials; application of animal husbandry practices that improve immunity and strengthen natural defences against disease, which primarily includes regular physical activity and, where appropriate, access to open areas and pastures; exclusion of artificially obtained polyploid animals; maintenance of the biological diversity of natural aquatic ecosystems, the long-term health of the aquatic environment and the quality of the surrounding aquatic and terrestrial ecosystems in fish farming production; feeding aquatic organisms with feed obtained from sustainable exploitation of fishery resources or with organic feed consisting of agricultural ingredients obtained from organic farming and from natural non-agricultural materials.

Organic production methods in plant and livestock production shall be applied from the beginning of the conversion period, which shall begin on the day of conclusion of the contract on the performance of organic production control between the producer and the authorized control organization. During the conversion period: organic production shall be carried out in accordance with the law governing organic production and the regulations adopted on the basis thereof; on an agricultural holding or production unit where organic production is partly carried out and partly the transition to organic production is in progress, organic products and products from the conversion period, as well as animals, shall be kept separately or their easy separation shall be enabled, and appropriate data shall be kept on the basis of which it can be determined how the products are separated. Organic production methods must be applied without interruption if the producer, who has concluded a contract on control and certification in organic production, changes the authorized control organization. Conventional production and organic production methods may be applied simultaneously on one agricultural holding (hereinafter referred to as: parallel production) if: in the case of crop production - different varieties of plants that are easily distinguishable from each other are grown on the agricultural holding, and in the case of livestock production - if animals of different species are raised, provided that in the case of fish and other aquatic animals, the same species of fish or other aquatic organisms may be raised, and in locations that are appropriately separated; ensure that production units, as well as animals and products used in organic production or obtained from organic production, are separated from those used in conventional production or obtained from conventional production, and on the condition that data are kept separately for organic and conventional production.

Exceptionally, if there are climatic, geographical and structural difficulties, parallel production may be carried out on an agricultural holding, as follows: in the case of the production of perennial plant species requiring a cultivation period of at least three years, varieties that cannot be easily distinguished may be grown, if the production forms part of a conversion plan that provides for the inclusion of the last production unit on which conventional production is applied in organic production in the shortest possible time (no longer than five years); if appropriate measures have been taken to spatially separate products from organic and conventional production; if the producer informs the authorized control body at least 48 hours before the harvest of each product; if, after the harvest, the producer informs the authorized control body of the quantity of products from the production units and of the measures applied for the separation of the products; if the producer submits an approved conversion plan to the authorized control body every year; in the case of agricultural areas used for research or education purposes, as well as in the case of the production of reproductive material, provided that the above conditions are met; if the meadow is used exclusively for grazing.

Exceptionally, if there are climatic, geographical and structural difficulties, parallel production may be carried out on an agricultural holding, in the case where livestock production is carried out for research or education purposes, with approved parallel production. In such cases, animals of the same species may be bred provided that: measures have been taken, of which the authorized control body has been notified in advance, to ensure the separation of livestock, livestock products, fertilizers and animal feed from each production unit; the producer informs the authorized control body in advance of each delivery or sale of animals or products of animal origin; the producer shall inform the authorized control organization of the quantity of organic products produced, stating the characteristics that enable the identification of the product and confirming that measures for product separation have been applied; when breeding bees, the conditions prescribed by this Regulation are met, except for the conditions relating to the location of the apiary, then parallel production may be carried out for pollination purposes. In the described cases, data on parallel production should be kept, and the product obtained in such a way is not considered an organic product.

The collection of organic and conventional products (e.g. harvesting, picking, collection of products of animal origin) may be carried out simultaneously, provided that appropriate measures are taken to prevent mixing or substitution of organic and conventional products and to ensure the identification of the products. The producer shall keep records of the date and time of collection, the method of collection, the quantity of products collected and the place of storage of those products.[20]

5. Methods of organic plant production

Methods of organic plant production include the selection of plant species and varieties, crop rotation, soil cultivation system, means and method of fertilization, system of maintaining soil fertility, method of controlling plant diseases, pests and weeds, method of collecting wild species from natural habitats, in accordance with the law regulating organic production. Agrotechnical measures applied in organic plant production should prevent or reduce to the minimum possible extent environmental pollution. Hydroponic production cannot be carried out in organic plant production.

a) Selection of plant species and varieties

In organic plant production, preference is given to plant species and varieties that are adapted to local agro-ecological conditions, i.e. autochthonous and domesticated varieties and varieties that are tolerant to disease and pest agents. In the production of organic products that are not reproductive material, reproductive material produced by organic production methods is used. For the purposes of these regulations, reproductive material produced by organic production methods is considered to be seed, i.e. vegetative reproductive material whose mother plant, i.e. parent plant, has been grown by organic production methods for at least one generation, or, in the case

[20] *Ibid.*, Art. 5–12.

of perennial plants, for at least two growing seasons. The availability of the above reproductive material can be determined on the basis of the reproductive material database, which is available to the public via the Ministry's website.

Plant species and varieties used as reproductive material in organic production are entered into the reproductive material database based on a notification from the reproductive material supplier. Any plant species and variety that is not entered into the reproductive material database is considered unavailable for organic production, i.e. non-existent on the market.

The above notification shall be accompanied by: a certificate that the reproductive material was produced using organic production methods in accordance with the regulations governing organic production; proof that the reproductive material placed on the market meets the general requirements of the regulations governing the recognition of varieties, i.e. the production and trade of seeds and planting material; data on the commercial and Latin name of the plant species and variety, the name of the supplier of reproductive material or its representative, and the contact address and telephone number, the area to which the supplier of reproductive material delivers reproductive material produced using organic production methods, and the estimated delivery time, then data on the name of the country in which the plant species or variety was tested and approved, the date from which the reproductive material produced using organic production methods will be available on the market, the name and code number of the authorized control organization that controls the supplier of reproductive material; a signed statement on the guarantee of timely notification in the event that it no longer has reproductive material used in organic production for that plant species and variety.

If the above conditions are met, the ministry enters the requested data into the reproductive material database. The supplier shall promptly inform the Ministry of the types and varieties of reproductive material produced by organic production methods that are no longer available on the market. The producer may use reproductive material for organic production obtained by sowing or vegetative propagation on his own property for his own needs, with the consent of the authorized control organization, provided that the plant production is carried out in accordance with the law governing organic production. In such a case, the producer shall inform the authorized control organization of the type and quantity of reproductive material that he intends to use for organic production in the following season.

Exceptionally, if there is no reproductive material produced using organic production methods on the market, in organic production, primarily reproductive material from the conversion period is used.

Also, in organic production, the use of reproductive material from conventional production may be approved before sowing, in accordance with the law regulating organic production, provided that this material has not been treated with plant protection products that are not permitted in organic production, if: there is no reproductive material produced using organic production methods on the market; the use of this material is justified for the purpose of conducting scientific research; it is an indigenous variety. In this case, proof shall be provided that the reproductive material has not been treated with plant protection products that are not permitted in organic production.

In the above case, when it comes to varieties, reproductive material from conventional production may be used in organic production, if the variety for which there is a need is not registered in the reproductive material database, if it can be proven that none of the registered varieties of the same species are suitable, and that the use of that variety is significant for production. If the user has ordered reproductive material in a timely manner, and no supplier, i.e. producer who places reproductive material on the market is able to deliver reproductive material before sowing or planting, it is considered that there is no reproductive material on the market that has been produced using organic production methods.

b) Crop rotation and soil cultivation system

Appropriate crop rotation and crop rotation in organic plant production ensures: maintaining soil fertility, reducing weed infestation and the number of pests and plant disease agents, reducing fertilizer leaching, maintaining the population of beneficial animal species and soil microbiological activity. Soil cultivation and plant cultivation in organic plant production are carried out in a manner that ensures the maintenance or increase of the organic matter content in the soil, the increase of soil stability and biodiversity, as well as the prevention of soil compaction and erosion. In organic plant production, agricultural machinery and equipment (machines for soil cultivation, sowing, fertilization, etc.) are used, which are cleaned and disinfected before use, if the machines are also used in conventional production and whenever necessary.

c) Means and method of fertilization and system for maintaining soil fertility

In organic plant production, plant nutrition agents and soil improvers may be used, to the extent necessary, if the nutritional needs of plants cannot be adequately met by applying measures for maintaining and improving soil fertility and activity. In organic plant production, plant nutrients and soil conditioners may be used, in accordance with the regulation governing plant nutrients and soil conditioners, which is given in Annex 1 - List of permitted plant nutrients and soil conditioners in organic production (hereinafter: Annex 1),[21] which is printed with these regulations and forms an integral part thereof. The total amount of fertilizer used in organic production, namely manure, dry manure and dehydrated poultry manure and composted animal excrement, including poultry manure, composted manure and liquid animal excrement, may not exceed 170 kg of nitrogen per ha of area per year, due to possible pollution of soil and water with nitrates. In order to redistribute surplus fertilizer from organic production, a producer applying organic production methods may transfer the use of that surplus fertilizer to another producer applying organic production methods on the basis of a written contract, in which case the permitted amount of fertilizer referred to in paragraph 3 of this Article shall be calculated on the basis of the areas of all production units covered by the contract. In organic plant production, appropriate preparations based on microorganisms may be used to improve the overall condition of the soil or the availability of nutrients in the soil or crops, appropriate preparations based on plants or microorganisms to activate compost, as well as biodynamic preparations. In organic plant production, the use of mineral nitrogen fertilizers is not permitted. For each production unit, data shall be kept on the area of the production unit (cadastral plot), the type, quantity and date of application of plant nutrients and soil conditioners in accordance with Annex 1 to these Regulations, which shall be documented by evidence of the need for their use.

Soil fertility and biological activity are maintained and increased by multi-year crop rotation, including legumes and other green manure, and by the application of manure or other organic fertilizer, preferably composted, originating from organic production.

---

[21] *Ibid.*, Annex 1.

d) Methods of pest, disease and weed control

Protection against pests, diseases and weeds is primarily carried out by the action of natural enemies, the selection of plant species and varieties, appropriate crop rotation, soil cultivation methods, the application of thermal processes, the use of allelopathic relationships and the establishment of protective plant belts, as well as other agrotechnical measures. If the application of the above measures does not adequately protect plants from pests, diseases and weeds, and it is determined that there is an immediate danger to plants, plant protection products may be used in accordance with the law, as provided in Annex 2 - List of active substances contained in plant protection products permitted for use in organic production, which is printed with these regulations and forms an integral part thereof (hereinafter: Annex 2).[22] If traps or sprayers are used in organic plant production (with the exception of pheromone sprayers containing plant protection products), it should be ensured that the release of substances into the environment is prevented and that contact of these substances with the plant species being grown is prevented. After use, the traps should be disposed of in a safe place. For each production unit, data on measures taken to control pests, diseases and weeds shall be kept, indicating the time of application and the agent used, which shall be documented by evidence of the need for their use.

e) Method of collecting wild species from natural habitats

Collection of wild plant species and their parts, which grow in natural habitats, forests and agricultural areas (hereinafter: wild species), shall be considered a method of organic plant production, if: these areas have not been treated with agents not permitted for use in organic production for a period of at least three years prior to collection; collection does not affect the stability of the natural habitat or the maintenance of species in the collection area; a protective zone has been established as isolation from sources of pollution and protection against potential contamination, when the authorized control organization determines that this is necessary; The collection shall be carried out in accordance with the regulation governing the protection of rare, rarefied and endemic endangered plant species and unprotected plant species. For each plant species, before the start of the collection of wild species, the following data shall be submitted to the authorized control organization within the deadline set by it: the common and Latin name of the wild species that are in the collection plan, as well as the date of the start and end of the collection period; a map

[22] *Ibid.*, Annex 2.

of the place where the wild species are collected with the boundaries of the collection area clearly marked; the name and surname of the person collecting the wild species; the location of potential sources of pollution in the collection area; the location of the purchase points; the location and full description of the storage facilities and other facilities used for the preparation and processing of the collected wild species.

In addition to the above data, the authorized control organization shall be provided with proof that the natural habitats have not been treated with other means for three years prior to collection, except for means permitted in organic production. For each wild species, during collection, data shall be kept on: the common and Latin name of the wild species, the area and place, i.e. the name of the area in which the collection is carried out, the method of collection, storage, processing and sale of the collected wild species, the date of collection, the quantity of collected wild species, as well as the names of the persons who collected the wild species from natural habitats.

f) Special rules in organic mushroom production

In organic mushroom production, substrates composed of manure and animal excrement originating from farms where organic production methods are applied may be used. In the event that such manure and animal excrement are not available, plant nutrients and soil conditioners from Annex 1 to these Regulations shall be used, provided that they do not exceed 25 percent of the weight of the total ingredients of the substrate, not including covering material and added water before composting: products of agricultural origin, except for the listed products from farms where organic production methods are applied; peat that has not been chemically treated; trees that, after being cut down, have not been treated with chemical products; inorganic fertilizers from Annex 1 to these Regulations, water and soil.[23]

6. Organic livestock production methods

Organic livestock production methods include the selection of animal species and breeds; the method of raising animals and animal breeding facilities; animal nutrition and health care; transport and slaughter of animals; the treatment of animals acquired from other farms; the method of collecting animal species from natural

[23] *Ibid.*, Art. 13–24.

habitats - all in accordance with the law governing organic production. Organic livestock production methods are applied on agricultural land over which the producer has the right of ownership, use, or lease, as well as on agricultural land covered by a cooperation agreement, which meets the needs of the animals in terms of available area for outdoor use and grazing.

a) Selection of animal species and breeds

Organic livestock production methods shall be applied to the following animal species: cattle (including buffalo and bison), horses, pigs, sheep, goats, poultry, bees. Specific organic livestock production methods shall be applied in the production of fish and other aquatic organisms listed in Annex 3 - Species of fish, crustaceans, echinoderms and molluscs in organic livestock production, which is printed with these regulations and forms an integral part thereof (hereinafter referred to as Annex 3);[24] as well as in the production of zooplankton, micro-crustaceans, rotifers, worms and other organisms used as food for fish and other aquatic organisms, adapted to those species.

When selecting the breeds or strains of animals to which organic livestock production methods are applied, the animals' ability to adapt to local conditions, their vitality and resistance to disease shall be taken into account. The breeds or strains of animals to which organic livestock production methods are applied shall be selected in such a way as to avoid certain diseases or health problems typical of breeds or strains used in intensive production, such as swine stress syndrome, pale, soft and watery meat, sudden death, spontaneous abortion and difficult birth requiring caesarean section. When selecting the breeds or strains of animals to which organic livestock production methods are applied, preference shall be given to indigenous breeds and strains. As regards bees, organic livestock production methods may only be applied to bees of the Apis mellifera carnica breed.[25]

b) Animal husbandry methods and animal husbandry facilities

In order to prevent overgrazing, compaction, erosion and soil pollution caused by animals or resulting from the transfer of their excrement, the number of animals in

[24] *Ibid.*, Annex 3.

[25] *Ibid.*, Art. 25–27.

organic livestock production should be limited. The total number of animals per unit of area in organic livestock production should ensure the production of up to 170 kg of nitrogen per year per ha of agricultural land, in accordance with Article 20, paragraph 3 of these Regulations. The maximum permitted number of animals per unit of area in organic livestock production is given in Annex 4 - Maximum permitted number of animals per hectare in organic livestock production, which is printed with these Regulations and forms an integral part thereof (hereinafter: Annex 4).[26] In organic livestock production, animals are, as a rule, reproduced naturally, but artificial insemination is also permitted, except for cloning and embryo transfer. In the case of artificial insemination referred to in paragraph 1 of this Article, the producer shall keep records thereof and inform the authorized control organization with which he has concluded a contract. Animal reproduction may not be promoted by hormones and hormonal preparations, except in cases related to veterinary treatment of individual animals, as regulated by these regulations.

In organic livestock production, animals shall not be tethered or isolated, except in the case of individual animals, for a limited period of time, and provided that there are reasons of a safety nature and reasons relating to the welfare and health of the animals. Exceptionally, in the case where it is necessary to ensure the establishment or maintenance of organic production on farms facing climatic, geographical or structural difficulties, cattle on small production units may be tethered, if their keeping in groups is not possible in accordance with their living needs and provided that they are provided with access to pastures during the grazing period, or access to open air at least twice a week, if grazing is not possible. In organic livestock production, in order to preserve the quality of the product and traditional production practices, the physical castration of animals may be carried out in accordance with the law governing animal welfare. In organic livestock production, procedures such as tail docking, tooth grinding, beak trimming and dehorning cannot be carried out systematically, except in individual cases prescribed by the law governing animal welfare. The above procedures are performed at the most appropriate age of the animal and by qualified personnel using anesthesia and analgesia, in order to minimize animal suffering. In organic livestock production, mutilation of animals, such as cutting the wings of a queen, cannot be carried out. In organic livestock production, preference is given to the breeding of slow-growing breeds/strains of poultry, in order to prevent the use of intensive breeding methods. In case the producer does not breed slow-growing breeds/strains, the minimum age of animals for slaughter is: 81 days for chickens; 150 days for capons; 49 days for Peking ducks; 70 days for female Muscovy ducks; 84 days for male

[26] *Ibid.*, Art. 4.

Muscovy ducks; 92 days for mallards; 94 days for guinea fowl; 140 days for turkeys and roasting geese; 100 days for turkeys.

In organic livestock production, animals are generally kept separately from animals from conventional production. For a limited period of time each year, animals from conventional production may use pastures where organic production methods are applied, provided that these animals originate from an extensive farming system and that no animals from organic production are present on that pasture at that time. Animals from organic production may graze on shared land provided that this land has not been treated with products not permitted in organic production for at least three years and that any animals from conventional production using that land originate from an extensive farming system. Whereby, no livestock product from animals from organic production during the use of this land shall be considered an organic product, unless it is proven that these animals were separated from animals from conventional production. When animals from organic production are moved from one pasture to another, these animals may graze on land from conventional production, provided that the intake of food from conventional production in the form of grass and other vegetation that the animals graze during this period should not exceed 10 percent of the total amount of food per year, calculated as dry matter in the food. Data and documentation on compliance with the requirements of this Article shall be kept as evidence. In organic livestock production, animals are raised in appropriate animal breeding facilities or in open air in areas where climatic conditions allow animals to stay outdoors. Buildings for raising animals in organic livestock production should have such insulation, heating and ventilation that allow air circulation, dust levels, temperature, relative humidity and gas concentrations to be maintained within limits that are not harmful to the animals and provide adequate natural ventilation and a sufficient amount of natural light. In buildings for raising animals, the number of animals raised should be appropriate, in order to allow the animals, depending on the species, breed, age, sex and size of the group, to have: sufficient space (sufficient space to stand in a natural position, the possibility to lie down normally, to turn around, to groom themselves, to assume all natural positions and to make all usual movements, such as stretching limbs and flapping wings); good living conditions; the possibility to satisfy specific needs and habits. The minimum area of indoor and outdoor spaces and other characteristics of facilities for breeding different species and categories of animals are given in Annex 5 - Minimum indoor and outdoor areas of accommodation for different species and categories of animals in organic livestock production, which is printed with these regulations and forms an integral part thereof (hereinafter: Annex 5).[27]

It is prescribed that facilities for breeding mammals should meet the following conditions: the floor should be adapted to the needs of the animals and not slippery, and at least half of its total area, as specified in Annex 5, should be solid flooring, which means that this part is not of a slatted structure; the area for lying down and resting should be of sufficient size, made of solid material that is not of a slatted structure, and should be comfortable, clean and dry, with a sufficiently dry litter made of straw or other suitable natural material, which may be improved or enriched with another mineral product from Annex 1 to these regulations. In addition to the above conditions, facilities for breeding pigs should have areas for movement that allow the animals to burrow and dig. Different substrates may be used for these purposes. In these facilities, calves may only be housed in individual boxes up to the age of one week; sows should be kept in groups, except in the last stages of pregnancy and during the lactation period; piglets should not be kept in cages or in slatted boxes. Calves that are sick may be housed in individual boxes for more than one week of age, or until cured, in order to prevent the transmission of the disease.

The regulation states that poultry facilities should meet the following conditions: the facility should be constructed in such a way that all poultry have easy access to open space; they should have openings of an appropriate size for entry and exit, and the total length of these openings should be at least 4 m per 100 m² of the area available to the poultry; at least one third of the area of the building should be built of solid material, i.e. not of a mesh or lattice structure and should be covered with a mat of material such as straw, sawdust or sand, and in the part that is of a lattice structure, the bars should be of a size and number appropriate to the type of poultry and the size of the flock, as given in Annex 5 to this Regulation; the maximum capacity of the building should be: 4,800 chickens, 3,000 laying hens, 5,200 guinea fowl, 4,000 female Muscovy or Peking ducks or 3,200 male Muscovy or Peking ducks, and 2,500 capons, geese or turkeys; the total usable area of the building for fattening poultry in any individual unit should not exceed 1,600 m²; in the case of buildings for laying hens, these buildings must have sufficient floor space accessible to the laying hens and suitable for the collection of droppings.

In the above-mentioned buildings, natural light may be supplemented by artificial light, in order to achieve a maximum of 16 hours of light per day, with an

[27] *Ibid.*, Art. 28-33; Annex 5.

uninterrupted night-time rest period of at least eight hours without artificial light. Poultry must not be kept in cages, and waterfowl must be provided with access to a stream, pond, lake or swimming pool, whenever weather and hygiene conditions permit, in order to meet the specific needs of this species.

In organic livestock production, animals must, whenever weather and soil conditions permit, have permanent access to open air, except where measures prohibiting the movement of animals have been introduced, which is done in accordance with the regulations governing the control of contagious animal diseases. The above-mentioned open-air areas may be partially covered with vegetation, and open-air areas for poultry should be mainly covered with vegetation, have protective structures and allow poultry free access to an appropriate number of drinkers and feeders. Herbivores, whenever weather conditions and soil conditions permit, should have access to pastures for grazing, except during the winter months, provided that they have access to pastures during the grazing period and that the winter housing regime allows the animals to move freely. Exceptionally, bulls older than one year should be provided with access to pasture or open-air areas throughout the year. In the event of specific management problems in organic livestock production, the final phase of fattening cattle may also be carried out indoors, provided that the duration of this phase does not exceed one fifth of their lifespan, i.e. no longer than three months. Poultry should be provided with access to open air for at least one third of their lives, and if, due to the measures introduced to prohibit the movement of animals, in accordance with the regulations governing the control of infectious animal diseases, poultry is kept indoors, it should be provided with constant access to sufficient quantities of roughage and other appropriate materials to meet its ethological needs.

The placement of apiaries, except during periods when there is no flowering or when the bee colonies are in a dormant phase, shall be carried out in such a way that, within a radius of 3 km from the apiary location, the sources of nectar and pollen originate primarily from plants from organic production, i.e. from wild plants and plants from conventional production that have been treated with products permitted in organic production and at a sufficient distance from sources that may lead to contamination of beekeeping products and endanger the health of bees. Beehives should be made primarily of natural materials that do not pose a risk of pollution of the environment or bee products, and only natural products, such as propolis, wax and vegetable oils, may be used in them, except when the use of other products is necessary for the prevention of diseases and veterinary treatment, which shall be carried out in accordance with these regulations. Beeswax used for the formation of a new bee colony, or for the replacement of beeswax during the conversion period, must come from organic production units. Exceptionally, in certain cases, beeswax from conventional production may be used, namely: if there is no beeswax from organic production on the market; if it is proven that the beeswax is not contaminated with substances not approved for organic production; provided that it comes from waxy honeycombs. Bees may not be destroyed for the purpose of obtaining beekeeping products. Honey extraction may not be carried out from combs containing brood, and chemical synthetic insect repellents may not be used during honey extraction.

c) Animal nutrition

In organic livestock production, animal feed is used that primarily comes from the agricultural holding on which the animals are raised and where organic production methods are applied, or from other agricultural holdings on which organic production methods are applied, and which are preferably from the same region. In organic livestock production, animals may not be kept or fed in a way that may cause anemia, nor may food be force-fed to them. In organic livestock production, young mammals should be fed primarily with mother's milk, and if that is not possible with raw milk, for at least: three months from birth for cattle (including buffalo and bison) and horses; 45 days from birth for sheep and goats; 40 days from birth for pigs. The diet of herbivores in organic livestock production should be based as much as possible on grazing, depending on the availability of pastures at different times of the year. In the diet of herbivores, except during seasonal grazing periods, at least 60 percent of the feed for the animals should come from the own production unit or, if this is not possible, from other agricultural holdings engaged in organic production and which, if possible, are from the same region, with data and documentation on the purchased feed for the animals being kept. At least 60 percent of the dry matter in the daily rations of herbivores should consist of roughage, fresh or dried, or silage and haylage, and in the case of animals used for milk production, at least 50 percent, during the first three months of lactation. In the feeding of pigs and poultry, at least 20 percent of the feed for animals should come from the own production unit or, if this is not possible, from other agricultural holdings engaged in organic production, and which are, if possible, from the same region, with data recording and documentation on the purchased feed for animals. In the feeding of pigs and poultry, roughage, fresh or dried, or silage should be added to the daily rations. In organic livestock production, up to 30 percent of the amount of feed in the ration may consist of feed for animals from the conversion period, unless the feed for animals from the conversion period is produced on the own agricultural holding. In that case, the share of such feed may be 100 percent. In the case where feed for animals from the conversion period is produced on the owner's own agricultural holding, on units where organic production has not been carried out in the last five years, and that feed is provided by grazing or harvesting pastures or plots with perennial forage plants in their first year of conversion, the share of such feed may be up to 20 percent. In the case where both feed for animals from the conversion period and feed for animals from the first year of conversion are used in the diet of animals, the total share of such feed should not exceed the specified percentage. The specified percentages are calculated on an annual basis and converted into dry matter in the feed.

Limited quantities of protein feed from conventional production may be used in the diet of pigs and poultry, only if producers cannot obtain protein feed exclusively from organic production. The highest percentage of protein feed from conventional production approved for a period of 12 months for the listed species is 5 percent of the total amount of feed at the annual level, calculated in dry matter of animal feed of agricultural origin. The producer shall keep records of the use of protein feed from conventional production.

In the event of loss of plants from animal feed, or in the event of restrictions being introduced, in particular as a result of extraordinary meteorological conditions, the occurrence of infectious diseases, contamination with toxic substances or as a result of fire, conventionally produced animal feed may be used, for a limited period of time, on a specific area and in relation to an individual producer, with the aim of maintaining or re-establishing organic production.

The following products may be used in animal nutrition and organic production: in the case of plant and animal feed from conventional production or other substances used in animal nutrition, covered by point 2 of Annex 6 - Animal feed, which is printed with these regulations and forms an integral part thereof (hereinafter: Annex 6),[28] provided that they are produced or processed without the use of chemical solvents; in the case of spices, herbs and molasses from organic production, provided that they are produced or processed without the use of chemical solvents, in a maximum amount of 1 percent in the daily ration of a particular species of animal, calculated on an annual basis and converted into dry matter in animal feed of agricultural origin; organic animal feed; animal feed of mineral origin given in point 1 of Annex 6 to these regulations; products of sustainable fisheries, provided that they are produced or processed without the use of chemical solvents, that their use is limited to non-herbivorous animals and that the use of fish protein hydrolysates is limited to young animals only; sea salt,

[28] *Ibid.*, Annex 6

coarse rock salt; animal feed additives given in Annex 7 Products, i.e. additives in the production of animal feed that may be used in animal nutrition, which is printed with these regulations and forms an integral part thereof (hereinafter: Annex 7).[29]

Growth stimulants and synthetic amino acids may not be used in the nutrition of animals in organic production. In organic livestock production, bees may be fed with honey, sugar syrup or sugar from organic production, if their survival is endangered due to weather conditions, and only in the period between the last honey extraction and 15 days before the start of the next period of filling the hives with nectar or honeydew. In addition to the above cases, bees may be fed with honey, sugar syrup or sugar from organic production during long-term extreme weather events or extraordinary circumstances that prevent the production of nectar or honeydew, in order to maintain or re-establish organic production. In the nutrition of bees in organic livestock production, bees in hives should be left with sufficient reserves of honey and pollen at the end of the production season, so that they can survive the winter.[30]

d) Animal health protection

In organic livestock production, the prevention of animal diseases is based on the selection of appropriate breeds and strains, appropriate breeding practices, feeding with high-quality animal feed, physical activity, an appropriate number of animals per unit area and keeping animals in hygienic conditions. Chemically synthesized allopathic veterinary medicinal products or antibiotics may not be used for preventive purposes in organic livestock production, except in cases related to veterinary treatment of individual animals, which shall be carried out in accordance with these regulations. Hormones and similar substances used for reproduction control or for other purposes (e.g. for induction and synchronization of estrus) may not be used in organic livestock production, except in cases related to veterinary treatment of individual animals. Substances that promote growth or production (including antibiotics, coccidiostats and other artificial growth promoters) may not be used in organic livestock production.

The building, animal housing facilities, equipment and utensils used in organic livestock production should be regularly cleaned and disinfected in a prescribed manner in order to prevent the spread of infection and the reproduction of disease-carrying organisms, and feces, urine and uneaten or spilled food should be regularly

[29] *Ibid.*, Annex 7.

[30] *Ibid.*, Art. 34–45.

removed in order to minimize unpleasant odors, as well as the presence of insects and rodents. For the cleaning and disinfection of buildings, animal housing facilities and utensils, the cleaning and disinfection agents for livestock production equipment and facilities given in Annex 8 - Cleaning and disinfection agents, which is printed with these regulations and forms an integral part thereof (hereinafter: Annex 8),[31] may be used. For the destruction of rodents, insects and other pests in animal housing facilities, the products given in Annex 2 to this Regulation may be used, whereby rodenticides may only be used in traps, and in accordance with the law governing plant protection products. Poultry breeding facilities and outlets, unless poultry is raised in turns or kept in outlets, or if it can move freely during the day, should be emptied at the end of each production cycle, so that the facilities can be cleaned and disinfected, and the outlets can be re-vegetated. The listed facilities and outlets should be empty for a period of at least four weeks, and data and documentation shall be kept as evidence of compliance with this period. If, despite the application of preventive measures, animals become ill or injured, their treatment should be started immediately, and if necessary - isolated in appropriate premises. In the treatment of animals in organic livestock production, preference shall be given to phytotherapeutic, homeopathic and immunological products, trace elements, animal feed of mineral origin from Annex 6 to these Regulations and feed additives from Annex 7 to these Regulations, provided that their therapeutic effect is effective in a particular species of animal and in the treatment of particular diseases. If the use of the products is not effective in the treatment of diseases or injuries, and if the treatment is necessary to avoid suffering or torture of the animal, chemically synthesized allopathic veterinary medicinal products or antibiotics may be used under the responsibility of a veterinarian.

If an animal or group of animals receives more than three treatments with chemically synthesized allopathic veterinary medicinal products or antibiotics within 12 months, or more than one treatment if their productive life cycle is shorter than one year (with the exception of vaccinations, parasite treatment and the implementation of mandatory animal health protection programmes), those animals or products obtained from them, in accordance with these regulations, cannot be sold as organic products and those animals must undergo a conversion period. After the use of chemically synthesized allopathic veterinary products, the withdrawal period of treated animals and their products shall be twice as long as the withdrawal period in conventional production, which is provided for in the instructions of the medicinal product manufacturer. In the event that the withdrawal period is not specified in the instructions of the medicinal product manufacturer, it shall be considered to last 48 hours. Data and

[31] *Ibid.*, чл. Art. 8.

documentation on the fulfillment of the conditions referred to in this Article shall be kept as evidence. When animals from conventional production are introduced into an agricultural holding where organic production methods are applied, then, depending on the circumstances, special measures may be applied for the diagnosis of infectious diseases, such as screening diagnostic tests or diagnostic measures for animals placed in quarantine. In organic beekeeping, apiaries may be disinfected mechanically by cleaning under steam or direct flame. To protect frames, hives and combs, primarily from pests, rodent control products may be used, which are used only in traps, as well as other appropriate products given in Annex 2 to this Regulation, and in accordance with the law regulating plant protection products. If, despite the application of preventive measures, the bees of the colony become ill or infected, their treatment should be started immediately, and if necessary, they should be isolated in separate apiaries, and veterinary medicines may be used in their treatment in accordance with the regulations governing the production and trade of veterinary medicines. In the case of infection with Varroa destructor, formic, lactic, acetic and oxalic acids are used, as well as menthol, thymol, eucalyptus oil or camphor. In that case, and only in order to stop the infection, it is possible to destroy the drone brood. If chemically synthesized allopathic products are used for treatment, the bee colonies should be placed in isolated apiaries during the treatment period and all wax should be replaced with wax obtained from organic production, after which the bee colonies should be included in a one-year conversion period in accordance with these regulations, except when the aforementioned products are used in the treatment.[32]

e) Transport and slaughter of animals

Animals in organic livestock production should be transported in such a way that the duration of transport is as short as possible. Allopathic tranquilizers may not be used when transporting animals in organic livestock production. Loading and unloading of animals in organic livestock production shall be carried out without the use of electrical stimulation to restrain the animals. When slaughtering animals in organic livestock production, suffering and mutilation should be reduced to the minimum possible extent.[33]

[32] *Ibid.*, Art. 46-51.

[33] *Ibid.*, Art. 52.

f) Treatment of animals acquired from other farms

In organic livestock production, as a rule, animals that are born and raised on an agricultural holding that applies organic production methods are used. If animals from organic production are not available in sufficient numbers, animals from conventional production may be imported to an agricultural holding that applies organic production methods for breeding purposes, as follows: 1. When the herd is first formed, young mammals may be imported, provided that they are raised in accordance with organic production methods immediately after the decision, with the condition that calves and foals are under six months old, lambs and kids under 60 days old, while piglets should weigh less than 35 kilograms; 2. When renewing a herd, adult females that have not been mated and male mammals may be introduced, provided that they are bred in accordance with organic production methods immediately upon arrival on the holding. Their number per year may not exceed: a maximum of 10 percent of the total number of female horses or cattle (including buffaloes and bison) and a maximum of 20 percent of female pigs, sheep and goats. In the case of holdings with fewer than ten horses or cattle, or fewer than five pigs, sheep or goats, a maximum of one female may be introduced per year. The number of female mammals that may be introduced into a herd may be increased by up to 40 percent in cases of: a significant expansion of the capacity of the production unit; a change in breed; the introduction of a new type of livestock production; the existence of a risk of cessation of breeding of certain breeds (in which case, female mammals that have been introduced into the herd may also be introduced, if they belong to that breed). If poultry from organic production is not available on the market, poultry from conventional production may be introduced into the agricultural holding, as follows: when the flock is first formed, provided that the chickens for egg and meat production are less than three days old; for the purpose of egg production, provided that the chickens for egg production are no more than 18 weeks old and have been bred in accordance with the provisions of this Regulation relating to nutrition and animal health protection. If animals from organic production are not available due to high mortality caused by health reasons, or due to an emergency situation, the herd or flock may be re-formed or restored with animals from conventional production, with the aim of maintaining or re-establishing organic production. Animals introduced from conventional production for breeding purposes and products obtained from them, as well as animals on the agricultural holding at the beginning of the conversion period, shall be considered organic after the end of the conversion period, in accordance with these regulations. In organic beekeeping, 10 percent of queens and swarms may be replaced annually with queens and swarms from conventional production, for the purpose of re-establishing apiaries, provided that the queens and swarms from conventional production are placed in hives with combs or honeycomb bases originating from organic production units. When organic bees are not available due to high mortality caused by health reasons or extraordinary circumstances, the apiary can be re-established or restored with conventional bees, with the aim of maintaining or re-establishing organic production.[34]

g) Special rules relating to fish and other aquatic organisms

Local species should be used in the organic production of fish and other aquatic organisms to obtain strains that are well adapted to production conditions, healthy, and capable of efficiently utilising available food sources.

When selecting species, care shall be taken to ensure that no significant harm is caused to wild species. Data and documentation on the origin and treatment of fish and other aquatic organisms shall be maintained as evidence.

Hormones and their derivatives shall not be used in the reproduction of fish and other aquatic organisms by organic production methods except for manual sorting. Artificial induction of polyploidy, artificial hybridisation, cloning, and the production of unisexual strains shall also not be used in the reproduction of fish and other aquatic organisms by organic production methods.

Organic production of fish and other aquatic organisms shall be based on rearing fry from parent stocks from agricultural holdings applying organic production methods. If fish and other aquatic organisms from organic production are not available, those of conventional output or natural habitats may be introduced into the agricultural holding for reproduction or genetic improvement. In such cases, animals introduced into the agricultural holding shall be reared according to organic production methods for at least three months before being used for breeding.

If no fry from organic production is available, fry from conventional production may be introduced into the agricultural holding for rearing purposes. In this case, organic production methods shall be applied during the last two-thirds of the production cycle, with a specified maximum share of fry originating from conventional production. It may also include the use of wild-caught fry in the event of a natural influx of larvae and fry of fish or crustaceans when stocking ponds or cages, and the farming of European glass eel, if a management plan exists for the specific site and if

[34] *Ibid.*, Art. 53, 54.

an artificial eel propagation system is not in place. Additionally, it includes the collection of eggs from natural habitats other than European eel eggs for farming in traditional extensive aquaculture in wetland areas, such as brackish water ponds, tidal areas, lagoons, and closed dikes, provided approved management measures are applied, and the fish are fed exclusively with food naturally available in the environment.

When rearing fingerlings in hatcheries and nurseries, parallel production may be carried out in some instances, provided that production units for organic production are physically clearly separated from conventional production and that a separate water distribution system is in place.

The rearing of fingerlings for consumption may be carried out in parallel production in some instances if the measures for separating the production units are based on the natural state, a separate water distribution system, the distance between the production units, the tidal flow and the upstream and downstream location of the organic production unit, and if different stages of production and handling take place in the production units.

Data and documentation demonstrating compliance with the requirements of this Article shall be maintained as evidence.

Fish and other aquatic organisms shall be handled carefully and as little as possible using appropriate equipment. Anaesthesia may be used when handling fish and other aquatic organisms to prevent stress and physical damage. Sorting should be limited to the minimum necessary to ensure the animals' welfare.

The stocking density of fish and other aquatic organisms for specific species or groups of species is provided in Annex 3 of this Regulation. The impact of stocking density on animal welfare – including damaged fins, other injuries, growth rate, behaviour, and health – shall be regularly monitored.

Organic production of fish and other aquatic organisms shall be carried out in locations not exposed to pollution caused by products or substances not permitted in organic production or by pollutants that may contaminate these products.

In cases of parallel production, separation measures for production units shall be based on the natural conditions, including separate water distribution systems, the distance between production units, tidal flow, and the upstream and downstream location of the organic production unit.

When establishing a new production process – specifically, production exceeding 20 tonnes of fish and other aquatic organisms per year – an environmental impact assessment shall be conducted for the production unit to determine both the conditions for production and the impact on the immediate surroundings of the production process.

For this purpose, a sustainable management plan shall be developed to update the production unit annually. The plan shall contain: Detailed data on the effects of the production process on the environment; an environmental monitoring procedure; a list of measures applied to minimise the negative impact on the aquatic and terrestrial environment, including the discharge of nutrients into the environment during the production cycle or annually; results of the control of the production process and the maintenance of technical equipment.

The sustainable management plan for the organic production of fish and other aquatic organisms should contain, among other things, defensive and preventive measures applied against predators.

Priority shall be given to using renewable energy sources and material recycling. A waste reduction schedule shall form part of the sustainable management plan from the start of production. Where possible, the use of waste heat shall be limited to energy derived from renewable sources.

Ponds used for the production of fish or other aquatic organisms shall be equipped with natural, biological, or mechanical filters for waste collection or shall utilise seaweed, shellfish, and/or algae that contribute to improving wastewater quality. The quality of the wastewater shall, where possible, be monitored.

Furthermore, cooperation shall be established between producers of adjacent production units when drafting a sustainable management plan.

In systems for producing fish and other aquatic organisms, the water's flow and physico-chemical parameters shall ensure the animals' health and welfare. The production system should be designed, located, and managed in such a way as to minimise the risk of animals escaping.

If fish or crustaceans escape from the production system, appropriate measures shall be taken to reduce the impact on the local ecosystem, including recapture where feasible. The producer shall keep records and documentation regarding such incidents.

The environment in which fish and other aquatic organisms are reared should be designed so that: fish and other aquatic organisms have sufficient space to live, in accordance with the specific needs of the species; fish and other aquatic organisms are reared in water of good quality, with sufficient oxygen content, in accordance with the specific needs of the species; fish and other aquatic organisms are reared at a temperature and under lighting conditions adapted to the specific needs of the species and the geographical location; in the case of freshwater fish farming, the type of bottom shall, as far as possible, correspond to the natural characteristics of the bottom; in the case of carp farming, the bottom shall consist of natural soil.

Artificial lighting may be used for the farming of fish and other aquatic organisms, but with the following restrictions: in order to respect the ethological needs, geographical conditions, and health of the animals farmed, the extension of daylight hours shall not exceed a maximum of 16 hours per day in total unless artificial lighting is used for reproductive purposes; in order to avoid changes in the intensity of lighting, which may have a disturbing effect on fish and other aquatic organisms, dimmed or background lighting shall be used.

A natural water source may be used for heating or cooling at all stages of production, and artificial heating or cooling of water is only permitted in the case of hatcheries and nurseries. If there is a health need, as well as during critical periods of reproduction or transport, water aeration may be carried out in the following cases: during an increase in temperature, a drop in atmospheric pressure, or pollution; during occasional procedures, such as sampling or sorting; in order to ensure the survival of fish and other aquatic organisms.

Water aeration may be carried out provided that renewable energy sources primarily power mechanical aerators and that data on each use of the aerator is kept in the production records.

Closed water recirculation systems in facilities for the production of fish and other aquatic organisms shall not be permitted except in hatcheries and nurseries or for the production of species used as feed for organically produced animals.

Farming units with a natural soil bottom shall meet specific conditions. Systems with flowing water shall be equipped to monitor and control both the flow and the quality of the water, including the water entering and leaving the system. Additionally, at least five per cent of the peripheral area of the unit shall be covered with natural vegetation.

In marine systems for producing fish and other aquatic organisms, the site shall be located in an area with sufficient water flow, adequate depth, and effective water mass exchange. This minimises the impact on the seabed and the surrounding water environment. The holding cage shall be designed, constructed, and maintained according to the environmental conditions it is exposed to.

Mussels and other aquatic organisms that feed by filtering water and are not influenced by human intervention shall be reared following regulations in waters of high ecological quality. All their nutritional needs shall be met naturally, except for juveniles reared in hatcheries or nurseries.

Organic mussel production may occur in the same aquatic environment as organic fish, seaweed, and snail production, i.e. within a polyculture system. It should be recorded in the sustainable management plan. It should occur in environments delimited by posts, floats, or other visible markers, within which mussels are appropriately contained using nets, cages, or similar structures arranged in a way that does not endanger the survival of other species. Where predator nets are used, it must be ensured that they do not pose a risk to birds that hunt in the water.

Mussels' stocking density in organic production shall not exceed that permitted in conventional production. Sorting, thinning, and regulating stocking density should be carried out in accordance with biomass production to ensure both animal welfare and high product quality.

Mussels may be treated with a lime solution once during the production cycle to control harmful organisms. If organisms causing pollution appear, they should be physically removed from the production unit and, if possible, returned to the sea far from the production unit. In organic mussel production, the cultivation of mussels on ropes is suitable, as well as other production methods given in Annex 3 to this Regulation (part relating to molluscs and echinoderms). Cultivation of mussels on the bottom is permitted only if it does not have a significant impact on the environment at the collection and growth sites. The environmental impact assessment should contain data on the survey, as well as a report from the exploitation area, which also represents a separate chapter of the sustainable management plan.

Oyster mussels may be cultivated in bags placed on scaffolds. Oysters shall be arranged in beds aligned with the tide to optimise production, and the structures used for oyster cultivation shall be placed to avoid the formation of dams along the coast. Oyster production shall be carried out in accordance with the criteria given in Annex 3 to this Regulation (part relating to molluscs and echinoderms).

Wild seeds from outside the production unit may be used in organic shellfish production, provided that records are kept regarding the time, place, and collection method. Such collection must not cause a significant environmental disturbance, and the seed must originate from an environment where there is little chance of survival through the winter or where the quantity of seed exceeds the needs of the ecosystem and the natural environment for seed collection.

Shellfish seed from conventional production may also be introduced into production units using organic production methods. For the clam *Crassostrea gigas*, priority shall be given to selective breeding to reduce hatching in the wild.

The following rules apply to the feeding of fish and other aquatic organisms reared using organic production methods:

Fish and other aquatic organisms shall be fed with feed that meets their nutritional needs, depending on their stage of development. The feed of plant origin shall come from organic production, while feed derived from fish and other aquatic animals shall come from the sustainable exploitation of fishery resources; growth promoters and synthetic amino acids shall not be used.

Fish and other aquatic organisms produced using organic production methods may be fed mineral nutrients listed in Annex 6 to this Regulation, animal feed additives, specific products used in animal nutrition, and processing aids listed in Annex 7 to this Regulation.

Feeding practices should prioritise animal health, high product quality – including a nutritional composition that ensures a high-quality final edible product – and minimal environmental impact. Carnivorous fish and other aquatic organisms may be fed with feed originating from fish and other aquatic organisms from organic production, fishmeal and fish oil from organic production, fishmeal, fish oil and ingredients obtained from fish intended for human consumption produced in sustainable aquaculture systems, as well as nutrients of plant and animal origin from organic production.

When feeding carnivorous fish and other aquatic organisms, a maximum of 60 per cent of the diet may consist of plant products from organic production.

In the daily diet of salmon and trout, astaxanthin primarily derived from organic production (e.g., organic crab production) may be used within the limits of their physiological needs. If astaxanthin from organic production is not available, natural sources of astaxanthin (e.g., *Phaffia yesat*) may be used.

During the growth phase, the following fish and other aquatic organisms are fed with natural food from ponds and lakes: species from the carp family (*Cyprinidae*) and other species grown in polyculture, including perch, pike, catfish, species of the genus *Coregonus*, and sturgeon; shrimp (*Penaeidae*) and freshwater crayfish (e.g., *Macrobrachium spp.*); tropical freshwater fish such as *Chanos chanos*, tilapia (*Oreochromis spp.*), and Siamese catfish (*Pangasius spp.*).

When natural food sources are insufficient, organic plant-based food primarily grown on the farm or seaweed may be used for nutrition. The producer must maintain data and documentation as evidence. If the diet of freshwater crayfish (*Macrobrachium spp.*) and Siamese catfish (*Pangasius spp.*) is supplemented with food from ponds and lakes, the following limits apply: a maximum of 10 per cent fishmeal or fish oil from sustainable fisheries for Siamese catfish and a maximum of 25 per cent fishmeal and 10 per cent fish oil from sustainable fisheries for freshwater crayfish. Organic cholesterol may be used as a supplement to their diet, or, if unavailable, conventional cholesterol obtained from wool, shellfish, or other sources may be used.

The transport of fish and other aquatic organisms produced using organic methods must be carried out to ensure the health and welfare of the animals in accordance with the law. In organic production, the suffering of animals, including during slaughter, should be minimised. Slaughter must be carried out using techniques that ensure momentary unconsciousness and insensitivity to pain, considering the size, species, and location of production when selecting the method of slaughter.

Disease prevention is based on maintaining optimal conditions in organic fish and other aquatic organism production. This is achieved through appropriate site selection, optimal farm management, and good breeding and management practices, which include regular cleaning and disinfection of facilities, the use of high-quality food, proper stocking density, and selection of species and strains. A health protection and disease prevention plan must be adopted and implemented, detailing biosecurity protection measures and other preventive measures.

To prevent diseases in organic production, the following rules apply: uneaten food, faeces, and dead animals should be immediately removed, if possible, to avoid significant environmental damage, to prevent attracting insects or rodents, and to minimise disease risk. Ultraviolet radiation and ozone may only be used in hatcheries and nurseries. For the biological control of ectoparasites, "cleaner" fish, freshwater, seawater, and sodium chloride solution should, as a rule, be used. The production system, equipment, and utensils must be regularly and properly cleaned and disinfected with the cleaning and disinfection agents listed in Annex 8 of this Regulation.

To prevent diseases, a resting period may be imposed in some instances, during which the cage or other structures used in production should be emptied, disinfected, and left empty before reusing. The resting period should be applied and documented after each production cycle in open-sea farming. This is also recommended for other production methods using tanks, ponds, and cages, although it is not mandatory for shellfish farming. Immunological veterinary medicinal products may be used to prevent and treat diseases in organic fish and other aquatic organisms.

If health problems arise despite the application of the above preventive measures, treatment of the disease should be started immediately in order to avoid animal suffering. In the treatment of fish and other aquatic organisms, preference shall be given to substances of plant, animal, or mineral origin in a homoeopathic solution, then plants and their extracts that do not have an anaesthetic effect, and then substances such as trace elements, metals, natural immunostimulants or permitted probiotics.

If the use of the listed substances is not effective in treating the disease, allopathic veterinary treatments, including antibiotics, may be used in the treatment, provided that their use is limited to a maximum of two treatments per year or one treatment per year in the case of animals whose production cycle is shorter than one year. This does not apply to vaccinations and mandatory animal health programs. In the event that allopathic veterinary treatments are not used, fish and other aquatic organisms cannot be placed on the market as organic products.

Regarding parasite treatment, the use of such treatments should be limited to two treatments per year or one treatment per year for production cycles shorter than 18 months unless the treatment is part of a mandatory health care program implemented in accordance with regulations.

After allopathic veterinary treatments, parasite treatments, and treatments are included in mandatory health care programs, the withdrawal period must be twice as long as the withdrawal period in conventional production, which is provided for in the instructions of the drug manufacturer. In the event that the withdrawal period is not specified in the instructions of the drug manufacturer, the withdrawal period is 48 hours. In the event that veterinary medicinal products are used in the treatment of fish and other aquatic organisms, the manufacturer must report the use of such products to the authorised control organisation before placing those products on the market as organic, and the treated animals must be clearly marked.[35]

7. Duration of the conversion period

The duration of the conversion period is determined based on the type of crop or livestock production.

a) Duration of the conversion period in organic crop production

The duration of the conversion period in organic crop production depends on the time of sowing or harvesting and is as follows: for annual crop species, it must be at least two years before sowing; for pasture and perennial forage crops, it must be at least two years before their use as feed for animals from organic production; and for perennial plant species that are not forage crops, it must be at least three years before the first harvest of organic products.

Exceptionally, at the request of a producer who has concluded a contract with an authorised control organisation and after the first control, the duration of the conversion period in organic plant production may be shortened or extended under the following conditions:

Shortening the conversion period: This is possible if the period from the moment of the submission of the producer's request is also taken into account, provided that the land parcels: Were subject to environmental protection measures, and applying those measures can guarantee that products not permitted in organic production have not been used on those parcels; were part of a natural or agricultural area that has not been treated with products not permitted in organic production for at least three years, with evidence to that effect.

Extension of the conversion period: This applies if the land has been contaminated with products not permitted in organic production.

Shortening the conversion period for already converted plots: The conversion period may be shortened for plots of land that have already been converted or are in

[35] *Ibid.*, Art. 55–75.

the process of conversion if these plots have been treated with a product not authorised in organic production under the following circumstances: As part of mandatory disease or pest control measures carried out by the competent authority; as part of scientific tests approved by the competent authority.

Shortening the conversion period for pastures and open areas used by non-herbivorous animals: The conversion period may be shortened to one year. If the land has not been treated with products not authorised in organic production during the previous year, this period may be shortened to six months.

If, during the decomposition of a product whose use is not permitted in organic production, a negligible level of residues remains in the soil at the end of the conversion period, or in the case of perennial plant species, the duration of the conversion period cannot be shortened. These products cannot be sold with labels referring to organic production methods.

b) Duration of the conversion period in organic livestock production

If animals from conventional production are introduced to an agricultural holding, the conversion period in organic livestock production will depend on the species. For horses and cattle, including buffaloes and bison for meat production, the conversion period is at least three-quarters of the animals' lifespan or 12 months. The conversion period is at least six months for small ruminants, pigs, and dairy animals. For poultry raised for meat production, introduced before the age of three days, the conversion period is at least ten weeks, while for poultry raised for egg production, the conversion period is at least six weeks.

When animals from conventional production are present on the agricultural holding, the conversion to organic livestock production applies to the entire production unit, including the livestock, pastures, and/or any land used for animal feed. If the animals are primarily fed with products from that unit, the total conversion period for the animals, pastures, and/or land may be reduced to 24 months.

In organic beekeeping, the conversion period is at least one year, except for renewing apiaries.

For aquaculture, the length of the conversion period depends on the type of facilities used for producing fish and other aquatic organisms and the existing animals.

The conversion period is 24 months for facilities that cannot be drained, cleaned, and disinfected; 12 months for facilities that can be drained or left standing; six months for facilities that can be drained, cleaned, and disinfected; and three months for facilities in an open system, including shellfish production facilities.

In exceptional cases, the length of the conversion period may be shortened if there is evidence that the facilities used for producing fish and other aquatic organisms were not exposed to products not permitted in organic production immediately before the start of the conversion period.[36]

8. Method of conducting control and measures to be determined in case of established irregularities

a) General rules for conducting control

Control in organic production is conducted in accordance with the law governing this area, ensuring objectivity, impartiality, the absence of conflicts of interest, efficiency, professionalism, and consistency. This control applies at all stages of organic production for all organic products, using appropriate control methods based on clearly documented procedures. The control process begins when the producer is included in organic production, specifically when the producer signs a contract with an authorised control organisation to perform control and certification in organic production.

The contract for the performance of control and certification in organic production should primarily contain information on the contracting parties and the type of organic production (including repackaging, distribution, and import of organic products), the number and area of the cadastral plot on which organic production is carried out and the cadastral municipality, the rights and obligations of the contracting parties, the conditions and manner of exercising the right to issue certificates, and the manner of protection in the event of non-fulfilment of contractual obligations. The authorised control organisation shall submit a photocopy of the concluded contract to the Ministry as soon as possible. The established control system should enable the traceability of each product at all stages of production, processing, and circulation, in accordance with the law governing food safety, in order to guarantee that the organic

---

[36] *Ibid.*, чл. 76–80.

product was produced or imported in accordance with the law governing organic production.

In the event of termination of the contract above, the authorised control organisation, at the request of another authorised control organisation with which the producer has concluded a new contract, shall, as soon as possible, submit proof of termination of the contract after the fulfilment of all contractual obligations, a list of irregularities and implemented and accepted corrective measures, a list of parcels and/or animals with their status, as well as other necessary documents and information on controlled production. The authorised control organisation with which the producer has concluded a new contract shall verify whether the producer has eliminated all irregularities listed in the list of irregularities of the previously authorised control organisation, whereby a producer to whom a suspension measure and/or sanctions have been imposed cannot be accepted. In the event of termination of the contract referred to in Article 82, paragraph 2 of this Regulation, the authorised control organisation shall notify the Ministry thereof without delay.

Before the control is carried out, an activity plan is adopted, the person who will carry out the control (hereinafter referred to as the controller) is determined, and working documentation is provided. In the control procedure, the same controller cannot control one producer for more than three years in a row.

For the first control, the producer must submit to the authorised control organisation a signed written statement containing information on: the production units, i.e. the activity in which it is engaged; practical measures to be taken at the level of the production unit, i.e. the activity in which it is involved, to ensure compliance with the rules of organic production; preventive measures to be taken to reduce the risk of contamination with conventional products or substances whose use in organic production is not permitted; and cleaning measures to be taken in storage facilities and throughout the production chain.

By means of the written statement referred to in paragraph 1 of this Article, the producer shall guarantee that the production and trade processes will be carried out in accordance with the rules of organic production. In the event of established irregularities or violations, specific measures shall be implemented to bring production into line with the prescribed rules or, in writing, to inform buyers in order to ensure that the labels referring to organic production are removed from these products. The said written declaration shall be certified by the authorised control organisation.

The producer shall inform the authorised control organisation of any change in the data provided. The authorised control organisation shall keep records of all producers subject to control and make the data on the name, surname, or name and address available to interested parties.

The producer should enable the authorised control organisation to carry out the control by providing: access to all parts of the production unit and premises, as well as invoices and relevant supporting documents; information that is objectively necessary for control, i.e. information on imported shipments in the case of importers and first recipients; and the results of its own production quality assurance programmes.

If the producer also has a production unit used for conventional production within the agricultural holding, the authorised control organisation should enable the control of that production unit and all premises and facilities within that production unit. If the producer purchases organic products for production purposes, it should check whether those products have a certificate issued in accordance with the law governing organic production.

Control in organic production includes complete physical control and control of the producer's documentation and the application of precautionary measures. The authorised control organisation carries out control at least once a year through control visits, primarily announced and unannounced if necessary, in accordance with the control plan for organic production.

Control in organic production may be carried out several times a year through additional control visits based on a general assessment of the risk of non-compliance with the rules of organic production for each producer separately. This considers the results of previous controls, i.e., the number and severity of identified irregularities, the difficulties observed in eliminating irregularities, the quantity and variety of products, and the risk of mixing organic products with products from conventional production.

During the general risk assessment, three levels of risk can be determined: low, medium, and high. At least 10 per cent of the total number of controls during one year are unannounced controls. The total number of controls during one year is the number of annual controls increased by the number of planned additional control visits.

During the inspection, the authorised control organisation, based on a risk assessment and in case of suspicion, takes samples of materials to determine the presence of substances whose use is not permitted in organic production and to check the production technology. The number of samples taken should be at least five per cent of the producers controlled by the authorised control organisation.

The inspection of the producer's documentation in the inspection procedure refers to examining the data it keeps on the activity in which the producer is engaged, as well as the inspection of financial documentation. To enable the inspection, the producer should allow the authorised control organisation access to the data it keeps and the financial documentation so that during the inspection, the authorised control organisation can, at any time, check: the supplier, i.e. the seller or exporter of organic products; types and quantities of organic products delivered to the production unit, raw materials and additives purchased, as well as their use, and, if necessary, the composition of the animal feed; types and quantities of organic products stored; types and quantities of organic products that have left the production unit of the recipient or buyer, except for end users, as well as the premises of the first recipient or storage facilities; types and quantities of organic products purchased and sold, as well as the supplier or seller or exporter and the buyer or recipient, in the case of producers who do not store or physically handle organic products.

The authorised control organisation may request other necessary data or documentation from the producer, on the basis of which it can be verified that there is a correct relationship between the input raw materials and the finished products. In the event that the manufacturer has multiple production units, the production units where conventional production is carried out and the input storage facilities, as well as data on sales, storage, and quantities of conventional products, are also subject to verification.

For products originating from conventional production that are used in organic production, the authorised control organisation checks the documentation and whether the seller's statement confirms that the products are not produced from GMOs or with the help of GMOs.

In the event that a producer involved in organic production concludes a contract with a third party to perform certain tasks related to organic production (hereinafter: subcontractor), control of organic output also covers these tasks. In this case, in addition to the data, the producer shall also submit to the authorised control organisation: a list of subcontractors with a description of their activities, specifying the names of the authorised control organisations with which these subcontractors have concluded a contract for control and certification; written consent of the subcontractors that their production unit is subject to control; and data on practical measures, including documentation, taken at the production unit to ensure the traceability of the products that the producer places on the market.

If different authorised control organisations carry out the control of the producer and his subcontractors, the written statement should also contain the consent of the producer and his subcontractors that these control organisations can exchange data on the controls performed.

A legal entity may conclude a cooperation agreement with other producers to carry out organic production. According to this Regulation, these producers constitute a group of cooperating producers (hereinafter: cooperating producers). Cooperating producers are producers who carry out the same and/or similar types of organic production on production units located in the same geographical area.

In the above case, for the purpose of carrying out control, the contract concluded by the producer with the authorised control organisation shall specifically state all cooperating producers with whom the producer has concluded a cooperation contract, the number and area of the cadastral plot on which organic production is carried out, the cadastral municipality, and the plant species, i.e. the number of heads/individuals.

The producer should establish its own internal control system so that the control of the implementation of organic production at that producer and at all cooperating producers is carried out by a professional authorised by him (hereinafter: internal controller) at least once a year, in accordance with this Regulation. The internal controller shall inform the authorised control organisation of the irregularities found at the given producer and each cooperating producer, as well as the corrective measures taken to eliminate the observed irregularities within a certain period.

The verification of the efficiency of the internal control system, the work of the internal controller in the field, and the control of a given producer and a selected number of cooperating producers are carried out by the controller of the authorised control organisation. In each case, the level of risk for cooperating producers is determined by taking into account the following risk factors: the number of years of organic production as a producer group; the number of new members annually included in the group; the mutual similarity of the production systems and crops of cooperating producers; and changes in the internal controller.

The number of cooperating producers that the authorised control organisation should visit during the annual control is determined based on the number of group members and the general risk assessment by multiplying the square root of the number of cooperating producers by 1 in the case of a low level of risk, by 1.2 in the case of a medium level of risk, or by 1.4 in the case of a high level of risk.

The authorised control organisation prepares a written report for each control performed, which the producer also signs in accordance with the law regulating organic production.

Based on the report on the controls performed and the request of the producer, the authorised control organisation decides on certification. It issues a certificate confirming that the product or production process complies with the law governing organic production and the regulations adopted on its basis.

The certificate should contain information on: the certificate number; the name and address, i.e. the name and seat of the producer and the activity it performs; the name, seat, and code number of the authorised control organisation; the type and assortment, as well as the quantity of the product; the type of production; the validity period; the date of control; the date and place of issue; as well as the seal and signature of the authorised person.

The authorised control organisation issues the producer a certificate in which all cooperating producers are listed (group certificate), and the sale of organic products is thus certified by that producer. The authorised control organisation publishes information on the issued certificates on its website.[37]

b) Special rules for the control of organic plant production and the collection of wild plant species from natural habitats

In organic plant production, as well as when collecting wild plant species from natural habitats, in addition to general data, the producer shall submit to the authorised control organisation data on: storage room, production room, plots, i.e. product collection areas, and if necessary, processing, i.e. packaging room; and the date of the last work on plots, i.e. product collection areas, that was not carried out in accordance with the rules of organic production.

When collecting wild plant species from natural habitats, the producer shall submit the prescribed data. In organic plant production, the producer shall inform the authorised control organisation in writing each year, within the deadline set by the

[37] *Ibid.*, Art. 81–93.

authorised control organisation, about the dynamics of crop production and submit a plot-by-plot overview.

In order to enable verification in organic plant production, the producer shall keep separate records of: the use of fertilisers (date of application, type and quantity of fertiliser, and plot on which the fertiliser was used); the use of plant protection products (date of application, type of plant protection product, reason, and method of application); the purchase of input materials – external inputs (date, type, and quantity of product, purchase documentation); and harvesting and reaping (date, plots, plant species and variety, and quantity of organic products or products from the conversion period).

c) Special rules for the control of organic livestock production

Animals in organic livestock production are identified using techniques adapted to each species, individually in the case of large mammals or individually or in rotations in the case of poultry and small mammals.

In addition to the above general data, in organic livestock production, the producer shall submit to the authorised control organisation data on animal breeding facilities, pastures, open spaces, etc., and if necessary, premises for storage, packaging, and processing of livestock, livestock products, raw materials, and input raw materials (external inputs); premises for storing manure; and practical measures to be taken, which include: a fertilisation plan approved by the authorised control organisation, together with a complete description of the areas under crops, a contract on the transfer of the use of surplus fertiliser (if concluded), as well as a management plan for the organic livestock production unit.

In order to enable verification, in organic livestock production, the producer shall keep in particular data on: the herd and flock management system; the origin and date of arrival, the conversion period, the identification mark, and the veterinary card for animals arriving on the holding; the age, number of animals, weight at slaughter, identification mark, and destination for animals leaving the holding; the loss of animals and the reasons for this; the type, including feed additives, the proportion of the different ingredients in the rations, the periods of access to open areas, and the periods of movement of animals (nomadic livestock) when certain restrictions are in force, in the case of animal feed; the date of treatment, diagnosis, dosage, type of medication, indications of the active pharmacological substance used in the treatment, and veterinary prescriptions for veterinary care containing the justification, as well as the withdrawal period.

All this information is necessary for the livestock product to be labelled and placed on the market as an organic product for the purpose of preventing and treating diseases and protecting health.

When veterinary medicinal products are used, the producer shall inform the authorised control organisation thereof before the animals or their products are placed on the market as organic products. Animals treated with veterinary medicinal products shall be labelled individually in the case of large animals or by turns or hives in the case of poultry, small animals, and bees.

When it comes to beekeeping, the producer shall submit to the authorised control organisation a map of an appropriate scale containing a list of the hives' locations, as well as appropriate evidence and, if necessary, appropriate analyses proving that the conditions laid down in the law governing organic production are met for the area where the bee swarms are located.

The producer shall keep records of: bee feeding (type and quantity of product, date of application, and information on which hives the product was used); type of product, active pharmacological substance, diagnosis, dosage, method of application, duration of treatment, and withdrawal period when veterinary medicinal products are used; measures taken during the extraction, processing, and storage of bee products; transfer of hive bases and honey transfer.

When veterinary medicinal products are used, the producer shall submit data on this to the authorised control organisation before the products can be placed on the market as organic products. In the case of the transfer of apiaries, the producer shall inform the authorised control organisation thereof, as well as of the deadline within which the transfer will be carried out.[38]

d) Special rules for the control of aquaculture

In the organic production of fish and other aquatic organisms, in addition to general data, the producer shall submit to the authorised control body: a full description

---

[38] *Ibid.*, чл. 94–100.

of the capacity for the production of fish and other aquatic organisms on land and at sea; a sustainable management plan, if necessary; an environmental impact assessment, if necessary; in the case of molluscs, a special chapter from the sustainable management plan on minimal environmental impact shall be submitted.

In order to enable verification in the organic production of fish and other aquatic organisms, the producer shall keep separate data on: the origin, the start of the conversion period for animals coming to the holding, as well as the date of arrival; the number of lots, age, weight, and destination of animals leaving the holding; lost fish; the type and quantity of fish food, and in the case of carp and related species, a documented record of the use of additional food; veterinary treatment with details of purpose, date of application, method of application, type of product, and withdrawal period; disease prevention measures with details of non-production period, cleaning, and water quality.

In the case of shellfish production, an inspector from an authorised control organisation will visit the plant before and during the production of maximum biomass.

e) Special rules for the control of units for the processing of products of plant and animal origin, products from aquatic animals

In the case where processing is carried out on a production unit for its own use or for the needs of another person, including special production units that are engaged in packaging and/or repackaging of such products and production units that are involved in labelling and/or re-labelling of such products, in addition to the data referred to in Article 85, paragraph 1 of the Regulation, the producer shall submit to the authorised control organisation a description of the facilities for the reception, processing, packaging, labelling, and storage of agricultural products before and after the procedures to which they are subjected, as well as a description of the procedure for transporting the products.

f) Special control rules for production units for the production of animal feed

For production units where animal feed is produced, in addition to the data, the manufacturer shall submit to the authorised control organisation: a description of the facilities for the reception, processing, and storage of products intended for animal nutrition before and after the procedures to which these products are subjected; a description of the facilities for the storage of other products used in the preparation of animal feed; a description of the facilities for the storage of cleaning and disinfection products; if necessary, a description of the compound of the animal feed that the manufacturer intends to produce, in accordance with the Regulation governing the production and declaration of animal feed, as well as the species or class of animal for which the animal feed is intended; if necessary, the name of the animal feed that the manufacturer intends to produce; data on the practical measures to be taken, which also contain indications (indicators) of the measures applied in the technological processing procedures.

Based on the information on the practical measures to be taken by the producer, the authorised control organisation shall carry out a general risk assessment for each processing unit and draw up a control plan which, depending on the potential risk, determines the minimum number of random samples.

In order to enable verification, for production units where organic feed is produced, the producer shall keep separate records on the origin, type, and quantity of feed ingredients, additives, sales, and finished products.[39]

g) Special rules for the control of imports of organic products

In the case of the import of organic products, the importer shall submit to the authorised control organisation data on the premises of the importer and their activities during the import, the place of entry of the products into the Republic of Serbia, as well as on the facilities and premises that they intend to use for the storage of imported organic products until their delivery to the first recipient. The first recipient of imported organic products, i.e. the importer when they are also the first recipient, shall submit data on the reception and storage facilities.

At the request of the authorised control organisation, data on the transport organisation from the exporter to the first recipient shall also be submitted. The importer and the first recipient of imported organic products shall keep separate data on the state of stocks and maintain separate financial documentation unless they operate within one unit. The importer shall, without delay, notify the authorised control

[39] *Ibid.*, Art. 105, 106.

organisation of each shipment of organic products imported into the Republic of Serbia and submit the documentation prescribed by a special regulation.[40]

h) Measures to be determined in the event of irregularities in the method of carrying out organic production

If, during the control, it is found that certain irregularities exist in the method of carrying out organic production, the authorised control organisation may impose corrective measures on the producer to bring the production into compliance with organic production standards in accordance with the law governing organic production. The corrective measures determined by the authorised control organisation depend on the severity and impact of the identified irregularity on the overall organic production process and the integrity of the organic product.

Irregularities in organic production can be classified as: Minor irregularities: These do not undermine the methods of organic production or the nature of the organic product; serious irregularities: These affect the application of organic production methods and may jeopardise the integrity of the organic products, but they do not have long-term consequences on the established organic production process; critical irregularities: These cannot be corrected and significantly damage the integrity of the organic product, mainly related to the use of prohibited substances in organic production, and may have long-term effects on the established production process.

A report is drawn up detailing the observed irregularities and proposed corrective measures, signed by both the authorised control organisation's controller and the producer at the conclusion of the control. A minor or serious irregularity that the producer fails to correct may escalate to a higher category of irregularity.

The corrective measures may include the following: Minor irregularities: A warning is issued with a deadline for the producer to correct the issue. This does not delay certification, and the correction will be checked during the next control or documented before the control; serious irregularities: A suspension measure is applied, which may delay certification until the irregularity is resolved, or the certificate/attestation may be suspended for a period not exceeding six months; critical irregularities: A sanction measure is applied, which could involve: Denial of the

[40] *Ibid.*, Art. 107, 108.

certificate; reduction of the scope of certification; exclusion of part of the production from certification; complete withdrawal of the certificate.

If the products are withdrawn from the market in order to remove the organic product label from the product and accompanying documentation, it is also possible to separate all affected products or lots from other organic products. In such a case, the authorised control organisation shall determine the deadline within which the producer must correct the identified irregularities, i.e. the deadline for implementing the ordered measures, and, in order to verify that the irregularity has been eliminated, may conduct additional control and/or take a sample if there is suspicion that an unauthorised product has been used in organic production. When the authorised control organisation determines the measures, the producer shall propose a way in which they will eliminate the identified irregularities, which the authorised control organisation shall accept. In such a case, the authorised control organisation shall notify the Ministry thereof in writing, without delay, providing the most important information on the corrective measures imposed.

If the producer refuses to act in accordance with certain corrective measures, persistently fails to comply with the rules of organic production, sells products as organic contrary to the decision of the authorised control organisation, does not enable control, or fails to provide any evidence or data on the production carried out, the authorised control organisation may exclude the producer from organic production, i.e. terminate the contract.

If several different control organisations participate in the organic production control procedure, the authorised control organisation that has identified irregularities in the manner of carrying out the production process should cooperate with other control organisations and/or competent control authorities from other countries that carry out activities in the field of organic production, in order to eliminate the observed irregularities.[41]

9. Content, form, and method of keeping records kept by the authorised control organisation, and content and form of summary records

The content of the records maintained by the authorised control organisation is outlined in Form No. 1 – Form of records kept by the authorised control organisation,

[41] *Ibid.*, Art. 109–113.

which is included with this Regulation and forms an integral part of them. These records are kept electronically. If the authorised control organisation uses a special computer program to record data about the producer and its production, the records may be maintained in a specific format, and the Ministry must be notified of this.

The authorised control organisation is also responsible for entering any changes in the data within the records. Data about the producer will be deleted from the records if the producer stops engaging in organic production, changes the control organisation, or if the producer and the authorised control organisation terminate the contract for control and certification.[42]

a) Content and form of summary records

The authorised control organisation is required to submit an annual report on organic production to the Ministry every year, no later than January 31 of the current year for the previous year. This report must contain data from the records in accordance with the law governing organic production. Based on the data from the annual reports submitted by the authorised control organisations, the Ministry is responsible for maintaining summary records on organic production, which should include all the data from the individual records. These summary records are maintained in Form No. 2 – Summary Records Form, which is included with this Regulation and forms an integral part of them.[43]

10. Technological processes of processing in organic production, ingredients, additives and auxiliary substances used in the processing of organic products, cleaning methods and cleaning agents for technological lines

a) Technological processes of processing in organic production and ingredients, additives and auxiliary substances used in the processing of organic products

In organic production, the technological processes of processing must be stabilize to ensure that food and feed are produced solely from agricultural ingredients obtained through organic production methods, except when such ingredients are unavailable in the market. The use of food and feed additives, ingredients from

[42] *Ibid.*, Art. 114–116.
[43] *Ibid.*, Art. 116.

conventional production with predominantly technological or sensory functions, as well as trace elements and processing aids, must be limited. These substances may only be used to the minimum extent necessary and only if there is a technological need, a specific nutritional need for food, or as a zootechnical need in the case of animal feed.

The processing of food and animal feed must be stabilized in such a way that it does not involve substances or methods that could mislead as to the true nature of the product. Processing processes should be carefully managed, with priority given to biological, mechanical, and physical methods. All technological processing operations, such as smoking and the use of ingredients, additives, and auxiliary substances used in the processing of products obtained by organic production methods, should be in accordance with the principles of good manufacturing practice. The preservation or conservation of organic products is not considered a technological processing operation.

In the processing process, the producer should establish and implement appropriate procedures based on the systematic identification of critical processing stages and update them as necessary in order to always guarantee that the products have been processed in accordance with the rules of organic production. Procedures based on the systematic identification of critical processing stages include: taking precautionary measures to avoid the risk of contamination with substances or products not permitted in organic production; implementing appropriate cleaning measures, monitoring their effectiveness, and keeping records of these procedures; and taking measures to prevent conventional products from being placed on the market with a label indicating organic production.

In cases where products that are not produced using organic production methods are prepared or stored in the same production unit, measures shall be taken to ensure that: the processing and treatment procedures are carried out continuously until the entire batch is processed, spatially or temporally separated from similar procedures carried out on products from conventional production and products from the conversion period; organic products are stored before and after processing and treatment, spatially or temporally separated from products from conventional production and products from the conversion period; records are kept of all procedures and processed quantities, with regular notification to the stabilize control; batches are identified and the mixing or substitution with products from conventional production is prevented; and processing and treatment procedures for organic products are carried out only after appropriate cleaning of the processing equipment.

b) Technological processing processes and ingredients, additives and auxiliary substances used for food production

Technological processes for processing raw materials obtained by organic production methods shall be carried out separately in time and/or space from all other technological processes in food production, including the processing of raw materials from the conversion period.

The following rules shall apply in the production of food obtained from raw materials produced by organic production methods: food shall be produced predominantly from ingredients of agricultural origin, with added water and table salt not being taken into account. Food shall contain only ingredients from organic production, and the same product shall not contain an ingredient of the same type of conventional output or from production from the conversion period. In the case of food produced from ingredients obtained during the conversion period, that food shall contain only one ingredient of agricultural origin.

In the production of food obtained from raw materials produced using organic production methods, the following may be used: drinking water and salt (with sodium chloride or potassium chloride as the main components), which are otherwise used in the food industry; additives, processing aids and other products given in Annex 9 – *Ingredients, additives and processing aids used in the processing of raw materials obtained using organic production methods, yeast and yeast products,* which is printed with this Regulation and forms an integral part thereof (hereinafter: Annex 9);[44] natural flavouring substances, nature-identical flavouring substances and natural flavouring preparations, in accordance with the regulations governing flavourings; minerals (including trace elements), vitamins and amino acids, only if a special regulation stipulates that their use is necessary in the food containing them; preparations of microorganisms and enzymes commonly used in food production, while enzymes used as additives may be used if they are listed in point 1 of Annex 9 to this Regulation; for the labelling of meat and eggshells, in accordance with the Regulation governing food additives, and for the traditional colouring of boiled eggs, which are placed on the market during a certain period of the year, natural colours and natural coating substances may be used, with the approval of the stabilized control, stabilization and notification of the Ministry; agricultural ingredients not obtained by organic production methods given in Annex 10 – *Permitted ingredients of agricultural origin not produced*

[44] *Ibid.*, Annex 9.

by organic production methods (hereinafter: Annex 10),[45] which is printed with this Regulation and forms an integral part thereof.

In the processing process, substances and processes that restore the properties of the product that were lost during the processing and storage of food and eliminate the consequences of negligence or that may mislead regarding the nature of these products may not be used. Exceptionally, special substances are used in accordance with this Regulation in the production of products from the wine and yeast sectors.

When calculating the weight of ingredients in a processed product, food additives from Annex 9 to this Regulation, which are marked with an asterisk in the column "E number," as well as yeasts and yeast products, are ingredients of agricultural origin, while products and substances are not considered to be ingredients of agricultural origin.

Organic grape and wine products are used in the production of stabilized wine in accordance with the regulations governing the production of stabilized wine.

If agricultural ingredients that are not obtained by organic production methods are not available on the market, agricultural ingredients that are not listed in Annex 10 to this Regulation may be temporarily used in the production of food, provided that the producer offers evidence of this. The use of agricultural ingredients is valid for no more than 12 months, but that period may be extended a maximum of three times.

Only organically produced substrates may be used for the production of yeast in organic production. Other permitted additives that may be used for the production of yeast and yeast products in organic production are the additives given in Annex 9 to this Regulation and the substances referred to in this Regulation.

In food and feed produced by organic production methods, organically produced yeast shall not be present together with yeast not obtained by organic production methods. If an extract or autolysate of yeast obtained by organic production methods is not available for the technological production process, a maximum of 5 per cent of extract or autolysate of yeast not obtained by organic production methods may be added, calculated on the dry matter content.

[45] *Ibid.*, Annex 10.

c) Technological processing processes and ingredients, additives and auxiliary substances used in the organic production of products from the wine sector

Technological processing processes in the organic production of products from the wine sector shall be carried out in accordance with the regulations governing the production, oenological processes and restrictions, protection of geographical indications, traditional expressions, declaration, and presentation of products from the wine sector unless otherwise prescribed by this Regulation.

Technological processing procedures for raw materials obtained by organic production methods shall be carried out separately in time and/or space from the same technological process in the production of products from the wine sector. Products from the wine sector shall be produced from organic grapes. Only products and auxiliary substances listed in Annex 11 – *Products and substances permitted for use in the organic production of products from the wine sector*, which is printed with this Regulation and forms an integral part thereof, may be used in the production of products from the wine sector.[46]

The following oenological practices and oenological agents may be used in the organic production of products from the wine sector: aeration or the use of gaseous oxygen (oxygenation); heat treatment, provided that the temperature does not exceed 75°C; centrifugation and filtration with or without inert filtration media, provided that the pore size is not less than 0.2 micrometres; use of carbon dioxide, argon or nitrogen, alone or in combination, to create an inert atmosphere and process the product in the absence of air; use of yeasts for the production of wine, dry or in suspension with wine; use of one or more permitted substances, with the possible addition of microcrystalline cellulose as an auxiliary substance, to stimulate the multiplication of yeasts; use of sulphur dioxide, potassium bisulphite or potassium metabisulphite, potassium disulphite or potassium pyrosulphite; treatment with charcoal for oenological use; clarification using one or more substances stabilized for oenological use; acidification using stabilized substances; deacidification using stabilized substances; use of preparations of yeast cell membranes; use of lactic bacteria; addition of L-ascorbic acid; in the case of dry wines, use of fresh, undiluted and healthy lees containing yeasts from a recent vinification of dry wine; effervescence by introducing nitrogen; addition of carbon dioxide; addition of citric acid to stabilize the wine; addition of tannins; addition of metatartaric acid; use of gum Arabic; use of potassium bitartrate to promote the precipitation of tartrate; use of copper citrate to eliminate deficiencies in the taste

[46] *Ibid.*, Annex 11.

or smell of the wine; use of oak chips in the production and ageing of wine, including the fermentation of fresh grapes and beyond; use of potassium alginate; reverse osmosis in order to increase the natural alcoholic strength by volume of the wine; use of ion exchange resins; blending; sweetening, whereby the must, concentrated must and rectified concentrated must must be organic products; enrichment (increasing the natural alcoholic strength by volume), provided that if sugar is used, the concentrated must or rectified concentrated must must be an organic product, and provided that the enrichment of the wine by partial concentration by cooling is not carried out.[47]

d) Technological processing processes and ingredients, additives and auxiliary substances used in the processing of animal feed

Animal feed is obtained from raw materials produced using organic production methods, which is carried out separately in time and/or space from conventional production.

In the production of animal feed obtained from raw materials produced using organic production methods, the following conditions must be met: animal feed must contain only ingredients originating from organic production, and the same product must not contain an ingredient of the same type of conventional output or from production during the conversion period. Raw materials used for the production of animal feed may not be treated with chemically synthetic solvents. Substances and processes that restore the properties of the product that were lost during the processing and storage of animal feed, that eliminate the consequences of negligence or that may mislead as to the nature of these products, may not be used. The listed products and substances may be used.

e) Cleaning methods and cleaning agents for technological lines[48]

If technological lines used in organic production are also used for processing products, i.e. raw materials from conventional production, these lines should be thoroughly cleaned and washed before use. The manufacturer, following the instructions of the authorised organisation, records the means and method of cleaning

[47] *Ibid.*, Art. 117–126.
[48] *Ibid.*, Art. 127.

technological lines (origin, supplier, date of application, type and name, quantity of cleaning and disinfection agents used).

11. Appearance of the label and national symbol

This Regulation stipulates that a domestic, unprocessed, certified organic product shall be marked with the label "ORGANIC PRODUCT," the national symbol, and the code/logo of the authorised control organisation that certified that product in accordance with the law governing organic production. If an unprocessed organic product contains multiple ingredients, that product shall be labelled as organic only if all ingredients are of organic origin.

A domestically processed certified organic product may be labelled with the label above, the national label, and the code/logo of the authorised control organisation that certified that product in the last stage of production only if it contains at least 95 per cent of the ingredients of agricultural origin that were produced in accordance with the law governing organic production.

Products of plant origin from the conversion period, which contain exclusively one ingredient of agricultural origin, may be labelled with the label "PRODUCT FROM THE CONVERSION PERIOD" and the code/logo of the authorised control organisation upon the expiration of a period of one year from the date of conclusion of the contract with the authorised control organisation, in accordance with the law governing organic production.

Advertising material, accompanying documentation, and labels shall also be marked in the above manner in accordance with the law governing organic production. Certified products, as well as their advertising material, accompanying documentation, and labels, may be marked with terms or abbreviations indicating the origin of the organic product (ecological, biological, organic, eco, bio, org., etc.), as well as with the private label of the producer or producer association, in accordance with the law governing organic production.

It is stipulated that the provisions of this Regulation do not apply to pet food and food for fur animals.[49]

---

[49] *Ibid.*, Annex 12.

a) Appearance of the label

The minimum diameter of the label circle is 20 mm, and in the case of small packages, the diameter of the circle can be reduced to a maximum of 9 mm. In case some reasons prevent an organic product from being marked in the above manner, the organic product may be marked with a black-and-white certified organic product mark on a black or white background.

The mark for a product from the conversion period is rectangular, framed by a black line 1 mm thick, with the text "PRODUCT FROM THE CONVERSION PERIOD." The mark shall be printed in Cyrillic script, Arial font, black on a white background. The minimum size of the mark referred to in paragraph 1 of this Article shall be 20 x 10 mm.[50]

The above marks shall be applied by printing or embossing the sign on a sticker or tag or on the front of the individual or collective packaging of the organic product, advertising material, accompanying documentation or label. They shall be legible, visible, and indelible.

In the event that processed products contain less than 95 per cent of ingredients of agricultural origin that are produced using organic production methods and in accordance with the prescribed technological processing procedures, the list of ingredients of that product shall indicate the ingredients that are of organic origin, as well as the total percentage of organic ingredients in relation to the total amount of ingredients of agricultural origin, by indicating the letters or percentages, which should be in the same colour, size, and font as when indicating other ingredients. The specified list of ingredients may contain the label "ORGANIC PRODUCT" and terms or abbreviations indicating the origin of the organic product (ecological, biological, organic, or eco, bio, org., etc.).

If the main ingredient of a processed product is of hunting or fishing origin, and if it contains other agricultural ingredients that are of organic origin, the list of ingredients of that product shall state the ingredients that are of organic origin, and the label, i.e. the stated terms, may be placed in the same field of vision as the name of the product.

In the case of processed animal feed containing at least 95% of the dry matter of nutrients obtained by organic production methods, a label shall be used for its labelling,

---

[50] *Ibid.*

which must be separate from the label regulated by the regulations on the declaration and labelling of animal feed. This label shall be of the same colour, format, or font that does not attract more attention than the description or name of the animal feed on the packaging. It shall be placed in the same field of vision as the label of the dry matter weight, which shall include: the percentage of individual ingredients obtained by organic production methods, the percentage of ingredients from the conversion period, the percentage of ingredients not included in the previous categories, and the total percentage of ingredients of agricultural origin. The label shall be accompanied by a list of the individual ingredients of the animal feed obtained by organic production methods, as well as a list of ingredients from the conversion period.

Animal feed produced in accordance with this Regulation, which contains variable amounts of raw materials obtained by organic production methods, i.e. raw materials from the conversion period and/or the products listed in this Regulation, may be marked with the following statement: "Can be used for organic production in accordance with the law governing organic production and the regulations adopted on its basis." In addition to the label, the listed products shall also bear a note stating the obligation to use animal feed in accordance with the law governing organic production and the regulations adopted on its basis.

Animals and products of animal origin produced during the conversion period shall not be marked with labels related to organic products nor with the national symbol.

b) Appearance of the national symbol

According to the Regulation, the national symbol is used to mark certified organic products that meet the conditions prescribed by the law governing organic production and the regulations adopted on its basis. This does not apply to products from the conversion period and certain prescribed organic products from this Regulation. The national symbol is a graphic combination of a plant – a tree, and soil, within which the national symbols of the Republic of Serbia are recognisable.

The appearance of the national emblem is given in Appendix 12 – *Appearance of the national emblem*, which is printed with this Regulation and forms an integral part thereof.

The same visual field with the national symbol should also contain the code number of the authorised control organisation, which contains: the acronym identifying the Republic of Serbia – "RS", in accordance with the international standard for two-letter country designations SRPS EN ISO 3166-1; the abbreviation for organic product, written in capital letters – "ORG"; a number consisting of three numerical digits, issued by the Ministry, in accordance with the law regulating organic production.

12. Method of storage and transport of organic products

a) Method of storage of organic products

It is stipulated that if, in addition to organic products, other agricultural products are stored in storage facilities, organic products shall be stored separately from other agricultural products, i.e. food or animal feed. The necessary measures must be taken to ensure the identification of the consignment of organic products and to prevent mixing or substitution with products from conventional production.

In addition, appropriate hygiene measures must be taken, and the effectiveness of these measures must be checked before storing organic products. Data on the actions taken must be kept. If organic products are packaged and labelled, they may be stored in the same warehouse premises as products from conventional production but in a separate section that is labelled in accordance with the law governing organic production.

During storage, care must be taken to ensure that organic products are clearly identifiable at all times. Plant nutrition and plant protection products, cleaning and disinfection products, and animal feed products that are not approved for use in organic production may not be stored in a production unit where organic production of plants and animals is carried out.

Allopathic veterinary medicinal products and antibiotics may be stored on an agricultural holding, provided that they have been prescribed by a veterinarian and in accordance with the provisions of this Regulation relating to disease prevention and veterinary treatment. These products must also be stored at a location under the control of an authorised control organisation.

In accordance with this Regulation, records of the storage of the products, as mentioned above, are also necessary.

b) Method of transportation of organic products

Certified organic products are transported separately from products from conventional production. However, if certified organic products are packaged and labelled, they may be transported together with products from conventional production in accordance with the law governing organic production. Organic products transported to other establishments, including wholesale and retail, must be packaged in appropriate containers, vehicles, or packaging that is sealed so that the contents cannot be tampered with without breaking or damaging the seal and label. Organic products must also be accompanied by a statement that includes the name and address (or name and registered office) of the producer, as well as the owner or seller of the product if they are not the producer; the name of the product, or food/feed, indicating that it is an organic product; the name or code number of the authorised control organisation and if necessary, the batch identification mark issued in the manner prescribed by this Regulation, which allows the batch to be linked to the producer's records. Certain data may be included in the accompanying document if it can be linked to the packaging, container, or means of transport of the organic product, provided that this document also contains data on the supplier and/or carrier.

The closure of the packaging, container, or vehicle is not required if there is direct transport between two producers involved in organic production, the products are accompanied by a document containing the relevant data, and both the sender and recipient maintain records of the transport, which are accessible to the authorised control organisation.

When transporting animal feed obtained by organic production methods to other production, processing, or storage facilities, the following conditions must be met: animal feed must be physically separated from in-conversion feed and feed from conventional production feed during transport; Vehicles/containers used for transporting conventional products can also be used for transporting organic animal feed, provided appropriate cleaning measures are carried out before the transport begins. The effectiveness of cleaning has been verified. Finished animal feed obtained by organic production methods must be transported physically or temporally separated from other finished products. During transport, records of the initial quantity of the product, as well as each quantity delivered in a single round, must be kept and submitted, upon request, to the authorised control organisation.

Live fish produced through organic production methods must be transported in suitable tanks with clean water that meets their physiological needs in terms of temperature and dissolved oxygen. Before transporting fish and fish products produced by organic production methods, tanks shall be thoroughly cleaned, disinfected, and rinsed. During transport, special precautions shall be taken to reduce stress, and the stocking density of the fish transported shall not reach a level harmful to the species. Data on transport conditions must be recorded.

Upon receipt of the organic product, it shall be checked whether the packaging or container in which the product is contained is properly closed and whether the labels with the specified data are present. Then, these data shall be recorded and compared.

III. Regulation on documentation submitted to the authorised control organisation for the purpose of issuing a certificate, as well as on the conditions and manner of sale of organic products

This Regulation specifies in more detail the documentation that must be submitted to the authorised control organisation for the purpose of issuing a certificate confirming that the certified organic product being imported has been produced in accordance with the law governing organic production and the Regulation adopted on the basis of that law (hereinafter: the certificate). The Regulation also outlines the conditions and manner of sale of organic products that may be sold together with products from conventional production only if they are packaged.

To issue the certificate, which will be valid for the same period as the shelf life of the imported product, the importer is required to apply to the import of organic products, a copy of the certificate for the organic products being imported, a copy of the invoice for the organic products being imported, and a copy of the certified single customs document for the imported organic products, issued in accordance with the law governing customs procedures. If the imported consignment contains several types of organic products grouped under a designation such as "fruit," "vegetables," etc., a document listing the individual organic products being imported must also be submitted to the control organisation.

The application for the import of organic products contains information on the name and address, i.e. the business name and seat, of the importer; the addresses of the facilities and premises used by the importer for the storage of imported organic products; the organic products being imported, namely: the name and quantity of the organic products, the country of origin of the organic products (the name of the country in which the organic product was produced), and the country from which the organic product was imported; the name and address, i.e. the business name and seat, of the exporter of the organic products; and the control body or control authority that has certified the organic products, which is on the list of the European Commission (*List of control bodies and control authorities in the organic production sector*, *List of recognised control bodies and control authorities for equivalence*, *List of recognised countries with equivalent organic production rules and control systems*).

Organic products may be sold together with products from conventional production only if they are packaged in packaging intended for the final consumer. If organic products are not packaged in the prescribed manner, they are sold separately from products from conventional production in a clearly marked place. Certified organic products must be sold separately from products from the conversion period to prevent any mixing.

Organic products placed on the market must be marked and declared in accordance with the special regulations governing organic production and the declaration, labelling, and advertising of food.

The producer is required to keep records of the sale of organic products for each month. These records must include the date of sale, the type and quantity of organic products sold (including aggregate quantities), and the data on the buyer of the organic products. However, the producer is not required to keep records of data on the end consumer in accordance with the law regulating food safety.

IV. Report on organic production in the Republic of Serbia for 2020

1. Market of organic products

a) Basic characteristics of the domestic market of organic products

Regarding the market for organic products in the Republic of Serbia, this report states that the domestic market for organic products has experienced more intense development over the last 5–7 years. This, as pointed out, is primarily visible in the example of the opening of specialised markets, first in Belgrade, and then in Novi Sad, as well as the opening of organic product stalls in certain markets, the formation of "organic corners" in the retail stores of all major retail chains, the establishment of specialised retail stores, online sales, and fairs, festivals and other events also contribute to market development. In addition, it is worth noting that in 2017, one of the large retail chains established a special chain of organic stores in Belgrade. From the initial three stores, it quickly expanded to seven stores. Additionally, it is noted that there are several chains of specialised stores in Belgrade, where you can also find products from small organic food producers.

However, it is concluded that, in addition to positive trends in the domestic market, which are most visible in large cities such as Belgrade and Novi Sad, the national market for organic products remains underdeveloped. The most important reasons cited include low purchasing power of consumers; a tiny percentage of the population having formed eating habits in which organic products occupy a significant place; an insufficient level of consumer confidence that organic products are produced according to prescribed methods; poor customer awareness of organic products, such as ignorance of organic product labels; still low availability of organic products in retail stores; as well as a completely low level of public awareness and knowledge both about organic products and about organic production itself and its numerous positive effects on environmental protection.

Nevertheless, this report concludes that consumers in Serbia, as well as in European countries, are becoming increasingly aware of the importance of the quality of the foods they consume and are interested in those foods that can provide them with safety regarding certain risks associated with food products. In doing so, it refers to research published in 2019 by the European Food Safety Agency (EFSA), which was carried out in 28 EU member states and subsequently in 5 pre-accession states (Albania, Montenegro, Serbia, North Macedonia, and Turkey). The research aimed to establish the overall interest of Europeans in food safety, including factors that influence decision-making regarding food choices, the main channels of information, awareness of food-related risks, changes in behaviour, and attitudes resulting from this knowledge.[51]

b) Value of production, export, import, and market value

According to this report, calculating the value of realised production of organic products, as well as the value of traded organic agricultural products, is crucial for policymakers in the field of organic agriculture. This is essential for monitoring the effects of agricultural policy measures and for setting target values.

For agricultural producers, processors, and participants in the agricultural product trade, knowledge of production and turnover indicators is crucial for determining marketing channels, promotional strategies, and production programs. Therefore, according to one study, a methodology is presented for calculating the value of organic production, export value, import value, and market value of organic products. These values were calculated for all types of organic products (vegetable, animal, fresh, and processed).

Given that agricultural production is seasonal, with significant price fluctuations during the season, the average prices and quantities of primary and processed plant and animal products during certain periods in 2019 were considered to calculate the value of organic production. According to valid data for 2019, 13,284 tons of organic products were exported from Serbia, and the value of exports was 29.75 million euros, of which fruit and fruit products accounted for over 28.7 million euros. Moreover, compared to the previous year, the total value of exports increased by approximately 2.3 million euros and, over the past four years, by almost 11 million euros.

As stated in this report, the calculation of data on the import of organic products reveals that the value of imports in 2019 was EUR 8,913,370.

For participants in the organic agriculture sector, the above data enables the planning of placement and production based on information on the market situation. The attached methodology in the study allows for the calculation of this parameter for the following years using a uniform methodology, ensuring the comparability of this

[51] Ivana Simić, *Organska proizvodnja u Srbiji 2020*, Nacionalno udruženje za razvoj organske proizvodnje Serbia Organika, Beograd, 2020, 47-48 (in Serbian).

parameter by year. As stated, the calculation of the domestic market value of organic products yields data indicating that the value of organic products on the domestic market is EUR 16,624,673.

2. Agricultural policy and rural development policy on the way to the EU

As stated in this report, improving the quality of life in rural areas and reducing poverty, along with achieving a more equal distribution of income and economic opportunities, are essential aspects of the sustainable development of rural areas that the Republic of Serbia strives for. In this sense, the sustainable development of rural regions requires the effective coordination of all policies that impact these areas and their resources. Alleviating problems for those coming from the agricultural sector and promoting a pattern of equality can significantly contribute to sustainable development.

Creating favourable conditions for living and working and retaining young people in rural areas by providing more attractive jobs and equal opportunities for their families is one of the primary goals of the policy aimed at ensuring more even regional development in the Republic of Serbia.

Part of the rural development policy focuses on the sustainable management of natural resources and environmental protection, as well as enhancing the quality of life in rural areas and reducing poverty. Sustainable resource management and environmental protection are the foundations for ensuring the long-term stability and quality of domestic agricultural production in the face of increasing risks in the global food market.

The specific character of the agricultural output is reflected in its high dependence on the volume and quality of limited natural resources. For this reason, the necessity of reactions to climate change, protection of agricultural land from permanent change of purpose, reduction of greenhouse gas emissions, protection of biodiversity and typical rural landscapes, and rational use of water resources, forests, and other natural potentials of rural areas requires the definition of a new incentive policy that will respect the concept of multifunctional agriculture.

With the first intergovernmental conference between the Republic of Serbia and the European Union, held on January 21, 2014, negotiations for Serbia's accession to the European Union formally began in Brussels. Following the opening of the negotiations, the phase of analytical review of the legislation (screening) commenced to assess the conformity of the Republic of Serbia's legislation with the European Union's legal acquis.

The established criteria for opening negotiations in Chapter 11 were that the Republic of Serbia present to the European Commission an Action Plan, which will serve as a basis for the transfer, application, and execution of the acquis of the European Union in agriculture and rural development; and that the Republic of Serbia has submitted a request to the European Commission for the entrustment of budget implementation tasks for IPARD II.

At the same time, as stated, the instrument for pre-accession assistance (IPA II) represents assistance to potential candidate countries for membership and to candidate countries for implementing the process of stabilisation and association in accordance with their specificities. The instrument's priority area, agriculture and rural development, is intended for candidate countries with the aim of preparing for the implementation and management of the EU's Common Agricultural Policy.[52]

3. SWOT analysis

In terms of SWOT analysis, this report states that the overview of the organic production sector in the Republic of Serbia identifies numerous positive trends and indicates that many participants in the value chain are striving to advance on the path of development. While multiple factors indicate excellent potential for the development of organic production, several essential obstacles and problems still stand in the way and must be overcome.

Some of the main obstacles are capital limitations at all levels of the value chain, weak organisation of participants along that chain, and low efficiency in production, processing, and marketing. On the other hand, Serbia is on the path to joining the European Union, and the implementation of the IPARD II program enables investors to finance more than 60% of their investments through IPARD funds.

The organic production sector is eagerly awaiting the implementation of agroecological and organic production measures through IPARD III, which will provide organic producers in plant production with the opportunity to achieve

---

[52] *Ibid.*, 52-64.

incentives significantly higher than the national ones contracted over a long-term period (five years).

Furthermore, it is stated that the SWOT analysis of the organic production sector in the Republic of Serbia, as presented in this report, leads to the conclusion that there are numerous advantages and that many opportunities in this sector are recognised. However, it also highlights that the sector is faced with multiple threats that will pose challenges to its further path of progress – challenges that must be addressed in order to fully exploit the advantages and great potential that the sector holds.[53]

4. Development of the private sector in less developed regions of Serbia

According to this report, the implementation of the PSD project, which spanned from 2017 to 2022, was carried out within the framework of German-Serbian development cooperation by the German Organisation for International Cooperation (GIZ). This project supported the development of competitiveness among micro, small, and medium-sized enterprises (MSMEs) in Serbia.

The project was active in different areas with the aim of strengthening public and private service providers. To achieve its goal, it focused on four key sectors: IT, the metal industry and mechanical engineering, sustainable agriculture, and tourism.

Results: Producers of organic products were enabled to optimise their business processes using a mobile application developed specifically for agriculture. Additionally, the project contributed to the strengthening of female entrepreneurship and the improvement of soft skills related to business communication for organic producers and others.[54]

[53] *Ibid.*

[54] *Ibid.*, 137.

V. Analysis of statistical data of organic production in the Republic of Serbia

1. Statistical data on the production of organic products in Serbia

In accordance with the report on organic production in the Republic of Serbia, it was determined that the area under organic production in the country has been increasing over time, along with the share of these areas in relation to the total arable land.

According to verifiable and valid recent data, an almost continuous increase in both categories was recorded from 2019 to 2023. The only exception was the year 2020, when a slight decrease occurred – likely due to the consequences of the global pandemic at that time.

Specifically, the area under organic production increased from approximately 21,265 hectares at the beginning of the observed period, with a share of 0.61%, to around 29,000 hectares by 2023, with a share of 0.83%.[55]

Table 1: Areas under organic production and share in the total cultivated area in Serbia (2019–2023)

Year	2019	2020	2021	2022	2023
Area under organic production (ha)	21265.50	19317.00	23527.40	25035.13	29002.31
The share of area under organic production in the total arable area (%)	0.61	0.55	0.68	No data	0.83

[55] *FiBL (2019): Area data on organic agriculture in Serbia 2019-2023. The Statistics.FiBL.org website maintained by the Research Institute of Organic Agriculture (FiBL), Frick, Switzerland. Available at https://statistics.fibl.org/europe/key-indicators-europe.html. Accessed May 8, 2025.*

Similar to the previously observed example, a permanent increase in the number of organic food producers in the Republic of Serbia is evident. Thus, in the given period, from 2019 to 2023, there was an increase from 373 producers of organic food at the beginning of the observed period to 525 producers at the end of this period, which amounts to an increase of about 29 per cent.

Chart 1: Growth in the number of organic food producers in Serbia (2019–2023)

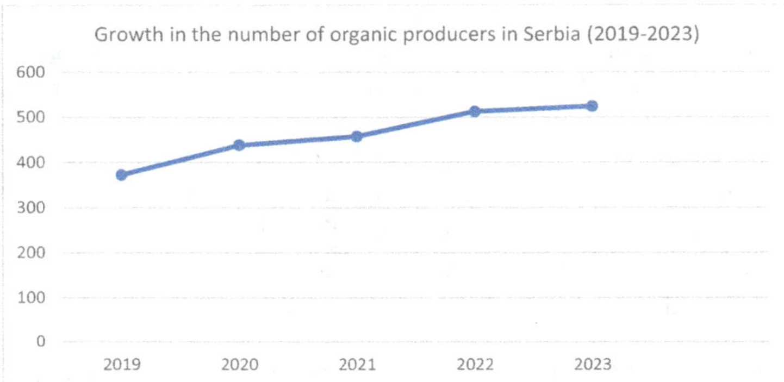

Data source:FiBL, op.cit.

2. Data on the export and import of organic products in foreign trade of Serbia (2015–2019)

Based on the available data on exports and imports for the period from 2015 to 2019, a nearly stable growth in the export of organic products can be observed. The export value increased from approximately 19.6 million euros at the beginning of this period to 29.7 million euros by the end, which represents an increase of about one third.

At the same time, although data on annual imports are available only for 2019, this report emphasises that calculating the market value of organic products is significant for agrarian policymakers in the field of organic agriculture. This calculation enables the monitoring of the effects of agrarian policy measures and the establishment of target values.

Table 2: Data on the export and import of organic products in foreign trade in Serbia (2015–2019)

Year	2015	2016	2017	2018	2019
Export (mil. EUR)	19,6	18,9	23,1	27,4	29,7
Import (mil. EUR)	No data	No data	No data	No data	8.9

Data source:Ivana Simić, Izveštaj o organskoj proizvodnji u Srbiji..., op.cit., 54-60

Chart 2: Growth of exports of organic products from the Republic of Serbia in the period 2015–2019 (mil. EUR)

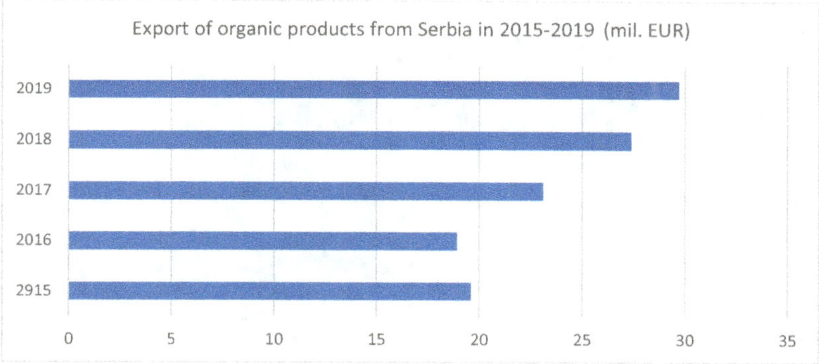

Data source:Ivana Simić, Izveštaj o organskoj proizvodnji u Srbiji..., op.cit., 54-60

3. Data on the market value of organic products in Serbia

The total value of the market of organic products in 2019 in the Republic of Serbia was calculated based on the calculated values of the total value of organic production in Serbia, the value of imported organic products and the value of exported organic products. By calculating the value of the domestic market for organic products in the Republic of Serbia for the year 2019, this report found that the total value of these products on the domestic market was approximately 16.6 million euros.

ANNEX 1

List of permitted plant nutrition products and soil improvements in organic production

Type and/or type

Mixed products or products containing only the materials listed below Description, compositional requirements, conditions of use

Manure products containing a mixture of animal excrement and plant material (animal bedding) are prohibited if they originate from intensive conventional production.

Dried manure and dehydrated poultry manure.

Composted animal excrement, including poultry manure and composted manure. Prohibited origin from intensive conventional production

Composted animal excrement, including poultry manure and composted manure. Prohibited origin from intensive conventional production

Liquid animal excrement (slurry). Use after controlled fermentation and/or appropriate dilution.

Prohibited origin from intensive conventional production.

Composted or fermented mixture of household waste. Products obtained from separated household waste, including vegetable and animal household waste, have been subjected to composting or anaerobic fermentation for the production of biogas. Use is permitted only if produced in approved, closed, and controlled systems, and the waste must be collected separately. Maximum concentrations in mg/kg of dry matter: cadmium 0.7; copper 70; nickel 25; lead 45; zinc 200; mercury 0.4; total chromium 70; chromium (VI) 0.

Peat Use is limited to horticulture (gardening, floriculture, tree cultivation and nurseries).

Mushroom production waste. The initial composition of the substrate may only contain products from this appendix.

Dead worm and insect remains (vermicompost).

Guano (seabird droppings).

A composted or fermented mixture of plant matter Product obtained from a mixture of plant matter that has been composted or subjected to anaerobic fermentation to produce biogas.

Digestate from biogas production containing animal by-products co-processed with plant or animal products listed in this Annex. Animal by-products (including those from wild animals) of Category 3 and digestive tract contents of Category 2 must not originate from intensive conventional agricultural production. Categories 2 and 3 are defined by specific regulations governing the classification of animal by-products. The handling of these materials must comply with the relevant regulatory framework. These products must not be used for the treatment of edible parts of plants.

Type and/or type

Mixed products or products containing only the materials listed below Description, compositional requirements, conditions of use.

Products or by-products of animal origin: blood meal, hoof meal, horn meal, bone meal or bone meal not containing gelatin, fish meal, meat meal, feather, hair and chiquette meal, wool, fur (1), hair, dairy products, hydrolysed proteins (2). 1) For fur: maximum concentration in mg/kg dry matter of chromium (VI): 0

(2) For hydrolysed proteins: these products must not be used to treat edible parts of plants

Products and by-products of plant origin for the production of fertilisers E.g. oilseed cake, coconut shell, malt waste.

Seaweeds and seaweed products. If they are directly obtained:

1. by physical processing, including dehydration, freezing and grinding
2. by extraction with water or aqueous solutions of acids and/or bases

3. by fermentation

Sawdust or wood scraps. Wood that has not been chemically treated after cutting.

Composted tree bark. Wood that has not been chemically treated after cutting.

Wood ash; From wood that has not been chemically treated after cutting

Finely ground raw phosphate. Conditions for use as specified in the special regulation governing the types of plant nutrition products and soil conditioners: cadmium content must be less than or equal to 90 mg/kg P_2O_5.

Aluminum-calcium phosphate. Conditions for use as specified in the special regulation governing the types of plant nutrition products and soil conditioners: cadmium content must be less than or equal to 90 mg/kg P_2O_5.

Basic slag. Conditions for use are specified in a special regulation governing the types of plant nutrition products and soil conditioners.

Raw potassium salts or kainite. Conditions for use are specified in a special regulation governing the types of plant nutrition products and soil conditioners.

Potassium sulfate, with possible magnesium salts. Potassium sulfate, with possible magnesium salts.

Ginger or ginger extracts Ammonia is not allowed.

Calcium carbonate (chalk, marl, limestone, Breton ameliorant, phosphate chalk). Natural origin only.

Mollusc waste. Only as a product of sustainable fisheries or organic aquaculture.

Magnesium and calcium carbonate. Only of natural origin e.g. magnesium chalk, ground magnesium, limestone.

Magnesium sulfate (Kaiserite). Only of natural origin.

Calcium chloride solution. Foliar treatment of apple trees, after determining calcium deficiency.

Calcium sulfate (gypsum). Conditions for use are specified in a special regulation governing the types of plant nutrition products and soil conditioners of natural origin only.

Industrial lime from sugar industry products.By-product in the production of sugar from sugar beet and cane.

Type and/or type

Mixed products or products containing only the materials listed below Description, compositional requirements, conditions of use

Industrial lime from vacuum salt production. By-product from the production of vacuum salt from rock salt mines.

Elemental sulfur. Conditions for use are specified in a special regulation governing the types of plant nutrition products and soil conditioners.

Trace elements. Conditions for use are specified in a special regulation governing the types of plant nutrition products and soil conditioners.

Sodium chloride. Rock salt only.

Stone flour and clay -

Leonardite (crude organic sediment rich in humic acids). Obtained as a by-product of mining

Chitin (a polysaccharide obtained from crustacean shells). Only if obtained from organic aquaculture or through the sustainable use of fish stocks, in accordance with special regulations.

Sediment rich in organic matter formed under anaerobic conditions by freshwater aquatic organisms (e.g. sapropel).Only organic sediment is a by-product obtained as a result of freshwater resource management or extraction from the surfaces of former freshwater ecosystems. Where applicable, extraction shall be carried out in a manner that minimises impact on the aquatic environment. Only sediment is obtained from sources that are not contaminated with pesticides, persistent organic pollutants, or petrol-like substances.

Maximum concentrations in mg/kg dry matter: cadmium 0.7; copper 70; nickel 25; lead 45; zinc 200; mercury 0.4; chromium (total) 70; chromium (VI) 0.

Xylitol. Obtained as a by-product of mining

Eggshells. Prohibited origin from intensive conventional production.

Hydrolysed proteins of plant origin.

Humic and fulvic acids. Only if obtained using inorganic salts/solutions, excluding ammonium salts, or if obtained by purifying drinking water.

ANNEX 2

List of active substances contained in plant protection products permitted for use in organic production

1. Substances of plant or animal origin

Name Description, compositional requirements, conditions of use
Allium sativum (garlic extract)
Azadirachtin extract from Azadirachta
indica (neem tree)
Beeswax For protection after pruning
COS-OGA
Hydrolysed proteins, except gelatin
Laminarin Seaweed is either grown using organic production methods or harvested sustainably in accordance with organic production principles.
Maltodextrin
Pheromones Only in traps and sprays
Vegetable oils All approved uses, except as herbicides.
Pyrethrin Only of plant origin
Quassia extracted from Quassia amaraInsecticide, repellent
Repellents of animal or plant origin repellent/sheep fat Edible parts of plants must not be treated. They must not be used to protect plants intended for the production of animal feed for goats and sheep.
Salix spp. Cortex (or willow bark extract)
Terpenes (eugenol, geraniol and thymol)

2. Basic substances

Basic substances Only those basic substances, as defined by the regulations governing plant protection products, may be used. These substances cannot be used as herbicides.

3. Microorganisms and substances produced by microorganisms

Name Description, compositional requirements, conditions of use
Microorganisms Microorganisms must not originate from GMOs
Spinosad
Cerevisan

4. Other substances not listed in Tables 1, 2 and 3

Name Description, compositional requirements, conditions of use

Aluminium silicate (kaolin) -
Calcium hydroxideFungicide: Only for fruit trees, including nurseries, to control Nectria galligena.
Carbon dioxide
Copper in the form of copper hydroxide, copper oxychloride, tribasic copper sulfate, copper oxide, and Bordeaux mixture.
Diammonium phosphate Only as an attractant in traps
Ethylene -
Fatty acids Approved uses, except as herbicide
Ferro (III) orthophosphate Products applied to the surface between cultivated plants
Hydrogen peroxide
Kieselgur (diatomaceous earth) -
Lime sulfur (calcium polysulfide) -
Paraffin oils -
Potassium and sodium hydrogen carbonate (potassium/sodium bicarbonate) -
Pyrethroids (deltamethrin or lambda-cyhalothrin only) Only in traps with special attractants; only against Bactrocera oleae and Ceratitis capitata Wied.
Quartz sand -
Sodium chloride Approved uses, except as herbicide
Sulfur-

Note: Unless otherwise specified in column two of this Annex, all substances shall be used according to the conditions of use specified in the List of Approved Substances.

ANNEX 3

Species of fish, crustaceans, echinoderms and molluscs in organic livestock production

I. Organic fish production

1. Organic production of freshwater trout
It refers to the following species of trout:
1) Brown trout (Salmo trutta);
2) Cutthroat trout (Oncorhynchus mykiss);
3) Brook char (Salvelinus fontinalis);
4) Salmon (Salmo salar);
5) Lake char (Salvelinus alpinus);
6) Grayling (Thymallus thymallus);
7) American lake char (Salvelinus namaycush);
8) Chum (Hucho hucho)

Production system The systems are supplied with food from open systems. The water flow should be sufficient to ensure at least 60% oxygen saturation, provide comfort for the fish, and eliminate the impact of wastewater.

Maximum holding density Species from the Salmonidae family not listed below: 15 kg/m³

Salmon: 20 kg/m³

Brook trout and California trout: 25 kg/m³

Lake Goldilocks: 25 kg/m³

2. Organic production of saltwater trout refers to the following species of trout:

1) salmon (Salmo salar);
2) brown trout (Salmo trutta);
3) cutthroat trout (Oncorhynchus mykiss)

Maximum holding density 10 kg/m3 in cages or nets

3. Organic production of cod (Gadus morhua) and other fish from the Gadidae family, sea bass (Dicentrarchus labrax), toothfish (Sparus aurata), hama (Argyrosomus regius), rumba (Psetta maxima [= Scopthalmus maximus]), pagara (Pagrus pagrus [=

Sparus pagrus]), Sciaenops ocellatus, and other fish from the suborder Sparidae and species of the genus Siganus.

Production system in open waters in cages and nets, with seawater velocity that ensures welfare, or in open systems on land.

Maximum holding density for all species except rumba: 15 kg/m³; For rumba: 25 kg/m³.

4. Organic production of sea bass, toothfish, hake, mullet (Liza, Mugil) and eels (Anguilla spp.) in inland tidal basins and coastal lagoons.

Holding system Traditional saltwater pools transformed into fish production units and similar tidal pools.

Production system A system with appropriate water mass renewal that ensures welfare.

At least 50% of the embankment should be covered with vegetation.

The submerged part should be purified.

Maximum holding density 4 kg/m3.

5. Organic production of sturgeon in freshwater

Refers to species from the sturgeon family (Acipenserideae).

Production system Each fish farming unit must maintain an appropriate water flow to support the health and welfare of the animals. The inflow and outflow of water should remain balanced, ensuring that the volume entering the system is equal to the volume being discharged.

Maximum holding density 30 kg/m3

6. Organic production of inland fish

It refers to species of the carp family (Cyprinidae) and other species cultivated in polyculture, including perch, pike, catfish, species of the genus Coregonus, and sturgeon.

Production system . In ponds that are periodically dried out and in lakes, fish farming should be oriented toward organic production, including the cultivation of crops along the shoreline. The areas designated for fish harvesting should include a sufficient number of clean-water bays to ensure fish comfort. After harvesting, fish must be transferred to clean water. Fertilisation using organic and mineral substances must comply with the guidelines set out in Annex 1 of this Regulation. The maximum allowable nitrogen application is 20 kg/ha. The use of synthetic agents to control aquatic plant growth in fish production areas is strictly prohibited. Natural vegetation must be preserved around the organic production unit to serve as a buffer zone, protecting it from adjacent land not managed under organic production methods. Polyculture production is permitted, provided that all criteria outlined in the specification for each additional species are fully met.

Yields: The total permitted fish production is 1,500 kg/ha per year.

7. Organic production of tropical freshwater fish

It refers to the following tropical freshwater fish:

1) Chanos chanos;
2) Tilapia (Oreochromis spp.);
3) Siamese catfish (Pangasius spp.).

Production systems Ponds and cages.

Maximum density Siamese catfish: 10 kg/m3

Tilapia: 20 kg/m3

II. Organic production of shrimp and freshwater crabs

It refers to shrimps (Penaeidae) and freshwater shrimps (Macrobrachium spp.)

Establishing producers units. The cultivation site should be chosen to minimise the unit's environmental impact. The pools must be constructed from natural clay. Mangroves must be preserved.

Conversion period Six months for the pool, related to the normal lifespan of farmed shrimp.

Origin of the parent flockAt least half of the parent stock should be domesticated after three years of farming. The remaining stock must consist of healthy wild organisms sourced from sustainable aquaculture systems. A mandatory health check for the first and second generations of the parent stock must be conducted before their introduction into farming.

Tentacle amputation: Not to be done.

Maximum stocking density and production restrictions. Seed: maximum 22 post-larvae per m2.

Maximum biomass: 240 g/m.

IIa. Organic production of freshwater crustaceans

Relates to the following species of freshwater crustaceans:

1) Astacus astacus;

2) Pacifastacus leniusculus.

Maximum density for small freshwater crayfish (< 20 mm): 100 individuals per m2.

For medium-sized freshwater crayfish (20–50 mm): 30 individuals per m2

For adult freshwater crayfish (> 50 mm): 10 individuals per m2

The condition is that the individuals have suitable shelters.

III. Organic production of molluscs and echinoderms

Production system: Long lines, rafts, bottom cultivation, net bags, cages, poles and other holding systems.

For mussel farming on rafts, the number of drop ropes should not exceed one drop rope per square metre of surface. The maximum permitted length of the drop rope is 20 m. The drop ropes should not be thinned during the production cycle. Splitting of the ropes may be carried out, but this does not increase the stocking density

ANNEX 4

Animal type and category: Maximum number of animals per ha (equivalent to 170 kg N/ha/year):

Horses over six months of age 2
Calves for fattening 5
Other cattle under one year old 5.
Male cattle 1–2 years old 3.3
Female cattle 1–2 years old 3.3
Male cattle two years old and older 2
Breeding heifers 2.5
Heifers for fattening 2.5
Dairy cows 2
Excreted dairy cows 2
Other categories of cows 2.5
Rabbits for breeding 100
Sheep 13.3
Goats 13.3
Piglet 74
Sows 6.5
Piglets for fattening 14
Other categories of pigs 14
Chickens 580
Brooding hens 230

ANNEX 5

Minimum internal and external accommodation area for different species and categories of animals in organic livestock production

1. cattle, sheep and pigs
Interior surfaces
(accessible to the animal) Outdoor areas
(exercise areas,
without pastures)
Minimum live weight
(kg) m2/head m2/head
Cattle and horses for breeding and fattening up to 100
up to 200
up to 300
over 350
1.5
2.5
4.0
5 or 1m2/100 kg
1.1
1.9
3
3.7 or 0.75 m2/100 kg
Dairy cows 6 4.5
Bulls 10 30
Sheep and goats 1.5 for a sheep, goat
0.35 for a kid sheep, goat 2.5
0.5
Breeding sows with piglets
up to 40 days 7.5 for a breeding sow 2.5
Fattening pigs up to 50
up to 85
up to 110
over 110 kg 0.8
1.1

1.3
1,5 0.6
0.8
1
1.2
Piglets Over 40 days and up to 30 kg 0.6 0.4
Breeding pigs 2.5 for a sow
6 for boars
If the space is used for natural
mating: 10 m2/boar 1.9
8
2. poultry
Interior surfaces
(accessible to the animal) External areas
(m2 available area/head)
Number of heads/m2 cm crossbar/neck nest
Carrier 6 18 litter per nest, or in the case of a shared nest, 120 cm2/head 4, ensuring that the limit
170 kg N/ha/year
is respected
poultry (in fixed facilities) 10 with a maximum of 21 kg
live weight/m2
20 (only for pearls)
20/5.000
20 (only for pearls) broilers and guinea fowl
4.5 ducks
10 turkeys
15 geese
For all the above species, the limit is 170 kg N/ha/year
Poultry fattening in a portable facility 161 in a portable facility,
with a maximum of 30 kg of live weight/m2 2.5, ensuring that the limit of 170 kg N/ha/year is respected 1 Only if the area of the movable structures does not exceed 150 m2

ANNEX 6

Animal food

1. Animal feed of mineral origin
   - Calcareous sea shells
   - Merl
   - Lithothamnion
   - Calcium gluconate
   - Calcium carbonate
   - Defluorinated monocalcium phosphate
   - Defluorinated dicalcium phosphate
   - Magnesium oxide (anhydrous magnesium)
   - Magnesium sulphate
   - Magnesium chloride
   - Magnesium carbonate
   - Calcium-magnesium phosphate
   - Magnesium phosphate
   - Monosodium phosphate
   - Calcium-sodium phosphate
   - Sodium chloride
   - Sodium bicarbonate
   - Sodium carbonate
   - Sodium sulphate
   - Potassium chloride

2. Other substances used in animal nutrition

Products/by-products obtained by fermentation with microorganisms whose cells are inactivated or dead:

- Saccharomyces cerevisiae
- Saccharomyces carlsbergiensis

ANNEX 7

Products or additives in the production of animal food that can be used in animal nutrition

Additives in animal nutrition must be approved in accordance with a specific regulation and may be used in accordance with the prescribed conditions of use unless otherwise provided for in this Annex

1. Technological additives

a) Preservatives

Identification number or functional group Substance, Description, terms of use

E 200 Sorbic acid -

E 236 Formic acid -

E 237 Sodium formate -

E 260 Acetic acid -

E 270 Lactic acid -

E 280 Propionic acid -

E 330 Citric acid -

b) Antioxidants

Identification number or functional group Substance Description, terms of use

1b306(i) Tocopherol extracts from vegetable oils -

1b306(ii) Tocopherol-rich extracts from vegetable oils (rich in delta-tocopherol) -

c) Emulsifiers, stabilisers, thickeners, gelling agents

Identification number or functional group Substance Description, terms of use

1c322 Lecithin Only from organic raw materials

It can only be used in aquaculture as animal feed

d) Binders and anti-caking agents

Identification number or functional group Substance Description, terms of use E 412 Guar gum

E 535 Sodium ferrocyanide Maximum dose of 20 mg/kg NaCl calculated as the ferrocyanide ion

E 551b Colloidal silicon dioxide

E 551c Kieselgur (diatomaceous earth, purified)
1m 558i Bentonite
E 559 Kaolin clays, asbestos-free
E 560 Natural mixture of steatite and chlorite
E 561 Vermiculite
E 562 Sepiolite
E 566 Natrolite/phonolite
1g568 Clinoptilolite of sedimentary origin
E 599 Perlite

e) Silage additives

Identification number or functional group Substance Description, terms of use

1k

1k236Enzymes, microorganisms

Formic acid It is only used in silage production if weather conditions do not allow for adequate fermentation. The use of formic and propionic acids and their sodium salts in silage production is only permitted when weather conditions do not allow for adequate fermentation.

1k237 Sodium formate
1k280 Propionic acid
1k281 Sodium propionate

2. Sensory additives

Identification number or functional group Substance Description, terms of use

2b Aromatic compounds Only extracts from agricultural products

Castanea sativa Mill.: extract chestnut

3. Nutritive additives

a) Vitamins, provitamins and well-defined chemical substances having a similar effect

Identification number or functional group Substance Description, terms of use

3a Vitamins and provitamins If synthetic, only those vitamins that are identical to those derived from agricultural products may be used for monogastric animals and aquaculture.

For ruminants, synthetic vitamins A, D, and E, which are identical to those obtained from agricultural products, may be used with the prior approval of the authorised control organisation. This approval will be based on an assessment of whether ruminants raised using organic production methods are able to obtain the necessary amounts of these vitamins through their regular diet.

3a920 Betaine, anhydride For monogastric animals only.
Only of natural origin and, if possible, organically produced
b) Microelements
Identification number
or functional group Substance Description, terms of use
E1 Iron
3b101 Iron(II) carbonate (siderite)
3b103 Iron(II) sulfate monohydrate
3b104 Iron(II) sulfate heptahydrate
3b201 Potassium iodide
3b202 Calcium iodate, anhydrous
3b203 Coated granular calcium iodate anhydrous
3b301 Cobalt (II) acetate tetrahydrate
3b302 Cobalt (II) carbonate
3b303 Cobalt (II) carbonate hydroxide (2:3) monohydrate
3b304 Coated granular cobalt (II) carbonate
3b305 Cobalt (II) sulfate heptahydrate
3b402
3b404
3b405 Copper(II) carbonate dihydroxy monohydrate
Copper(II) oxide
Copper(II) sulfate pentahydrate
3b409 Dicopper chloride trihydroxide (TBCC)
3b 502
3b 503 Manganese(II) oxide
Manganese sulfate, monohydrate
3b603
3b604
3b605
3b609 Zinc oxide
Zinc sulfate heptahydrate
Zinc sulfate monohydrate
Zinc Chloride Hydroxy Monohydrate (TBZC)
3b701 Sodium molybdate dihydrate
3b801 Sodium selenite
3b810, 3b811, 3b812, 3b813 and 3b817

Selenised inactivated yeast

4. Zootechnical additives

Identification number or functional group Substance Description, terms of use 4a, 4b, 4c and 4d Enzymes and microorganisms are in the category of "zootechnical additives."

ANNEX 8

Cleaning and disinfection products

1. Products for cleaning and disinfecting equipment and facilities for the production of fish and other aquatic organisms

1.1. In accordance with national regulations, and in particular with the regulation governing the use of biocidal products, products for cleaning and disinfecting equipment and facilities for the production of fish and other aquatic organisms used when there are no fish or other aquatic organisms in them may contain the following active substances:

- ozone
- sodium hypochlorite
- calcium hypochlorite
- calcium hydroxide
- calcium oxide
- caustic soda
- alcohol
- copper sulphate
- potassium permanganate
- tea seed mixture prepared from wild camellia seeds (use limited to shrimp production)
- a mixture of potassium peroxomonosulfate and sodium chloride for the production of hypochlorous acid.

1.2. In accordance with national inventories, and in particular with the regulation governing the use of biocidal products, products for cleaning and disinfection of equipment and facilities for the production of fish and other aquatic organisms and seaweed used in the presence of fish and other aquatic organisms may contain the following active substances:

- limestone (calcium carbonate) for pH control
- dolomite for pH correction (use limited to shrimp production)
- sodium chloride
- hydrogen peroxide

- sodium percarbonate
- organic acids (acetic acid, lactic acid, citric acid)
- humic acid
- peroxyacetic acid
- peracetic and peroctanoic acids
- iodoform (for caviar only).

2. Cleaning and disinfection agents for livestock production equipment and facilities are:

- potassium and sodium soap
- water and steam
- milk of lime
- lime
- quicklime (calcium oxide)
- sodium hypochlorite (e.g. liquid bleach)
- caustic soda
- caustic potash
- hydrogen peroxide
- natural plant essences
- citric, persic, formic, lactic, oxalic and acetic acids
- alcohol
- nitric acid (dairy equipment)
- phosphoric acid (dairy equipment)
- formaldehyde
- products for cleaning and disinfecting udders and milking equipment
- sodium carbonate

ANNEX 9

Ingredients, additives and auxiliary substances used in the processing of raw materials obtained by organic production methods, yeast and yeast products

1. Additives, including carriers

E number Name Food preparation Special conditions
Of plant origin Animal origin
E 153 Charcoal X Ashy goat cheese
Morbier cheese
E 160b (*) Anatto
Bixin, Norbixin X Red Gloucester Cheese
Double Gloucester Cheese
Cheddar Cheese
Mimolette Cheese
E 170 Calcium carbonate X X Prohibited use as a colouring agent or for calcium fortification of foods
E 220 Sulfur dioxide X X (only for mead) In fruit wines (**) and mead without added sugar: 100 mg/l (***)
E 223 Sodium metabisulfite X Crabs
E 250 or X For meat products (1)
May be used if the Ministry is provided with proof that no other technological solution exists that would offer the same guarantees and/or maintain the specific characteristics of the product. Not to be used in combination with E 250.
Indicative input amount (expressed as NaNO3): 80 mg/kg
Maximum residual amount (expressed as NaNO3): 50 mg/kg
E 252 Potassium nitrate X For meat products meat products (1)
May be used if the Ministry is provided with proof that there is no other technological solution that would provide the same
guarantees and/or allows the maintenance of the specific characteristics of the product. Not in
combination with E 250
Indicative input amount expressed as NaNO3:

80 mg/kg
Maximum residual amount expressed as NaNO3: 50 mg/kg
E 270 Lactic acid X X -
E 290 Carbon dioxide X X -
E 296 Malic acid X -
E 300 Ascorbic acid X X Meat products (1)
E 301 Sodium ascorbate X Meat products (1) related to nitrates and nitrites
E 306(*) High tocopherol extract X X Antioxidant
E 322(*) Lecithin X X Milk products (1)
Only if obtained from raw materials from organic production, and from 1 January 2022 from organic production
E 325 Sodium lactate X Meat and dairy products
E 330 Citric acid X X
E 331 Sodium citrate X X
E number Name Food preparation Special conditions
Plant origin Animal origin
E 333 Calcium citrate X
E 334 Tartaric acid (L (+)–) X X (only for mead) Food products of animal origin: mead
E 335 Sodium tartrate X
E 336 Potassium tartrate X X
E 341 (i) Monocalcium phosphates X X Raising agent for self-rising flour
E 392(*) Rosemary extract X X Obtained from organically grown rosemary
E 400 Alginic acid X X Dairy products (1)
E 401 Sodium alginate X X Dairy products (1)
E 402 Potassium alginate X X Dairy products (1)
E 406 Agar X X Dairy and meat products, meat products (1)
E 407 Carrageenan X X Dairy products (1)
E 410(*) Carob seed gum (Carob gum) X X Only if obtained from organic production –
applies from 1 January 2022
E 412(*) Guar gum X X Only if obtained from organic production – applies from 1 January 2022
E 414(*) Gum Arabic X X Only if obtained from organic production applies from 1 January 2022

E 415 Xanthan gum X X Only if obtained from organic production – applies from 1 January 2022
E 417 Tara rubber powder X X Thickener
Only if obtained from organic production – applies from 1 January 2022
E 418 Gellan gum X X Only the form with high acyl content
Only if obtained from organic production – applies from 1 January 2022
E 422 Glycerol X Only of plant origin
For plant extracts and flavours
Only if obtained from organic production – applies from 1 January 2022
E 440(*) (i) PectinX X Dairy products (1)
E 464 cellulose X X Encapsulating material for the production of capsules
E 500 Sodium carbonatesX X
E 501 Potassium carbonates X
E 503 Ammonium carbonates
E 504 Magnesium carbonates X
E 509 Calcium chloride X Milk coagulation
E 516 Calcium sulfate X Carrier
E 524 Sodium hydroxide X Surface treatment of "Laugengebäck" and regulation of acidity in aromas of organic origin
E 551 Silicon dioxide X X For herbs and spices in dry powder form
Aroma and Propyllis
E 553b Talc X X As a coating agent for meat products
E 901 Beeswax X As a glazing agent in confectionery only
Organically produced beeswax.
E 903 Carnauba wax X As a glazing agent for desserts
As a mitigating method of mandatory extreme cold treatment of fruit as a quarantine measure against harmful organisms
Only if obtained from organically produced raw materials
E 938 ArgonX X
E 939 Helium X X
E 941 Nitrogen X X
E 948 Oxygen X X
E 968 Erythritol X X Only if obtained from organic production without the use of ion exchange technology
Notes:
(1) The restriction applies only to products of animal origin.

(*) Food additives marked with an asterisk in column E number are counted as ingredients of agricultural origin for Article 120, paragraph 4 of this Regulation
(**) In this context, "fruit wine" is defined as wine made from fruits other than grapes (including cider wine and sparkling pear brandy).
(***) Maximum concentrations available from all sources expressed as SO2 in mg/l.

2. Permitted aid substances and other products used for the processing of raw materials

Obtained by organic production methods

Name Plant-based food preparation Preparation of food of animal origin Special conditions

Water X X Drinking water – regulated by positive regulations

Calcium chloride X Coagulant

Calcium carbonate X

Calcium hydroxideX

Calcium sulfate X Coagulant

Magnesium chloride (or nigari) X Coagulant

Potassium carbonate X Grape drying

Sodium carbonate X X

Lactic acid X For regulating the pH value in brine used in cheese production (1)

L(+) Lactic acid
obtained by fermentation X For the preparation of plant protein extracts

Citric acid X X

Sodium hydroxide X Sugar production

Oil production, excluding olive oil production, for the preparation of vegetable protein extracts

Sulfuric acidX X Gelatin production (1)

Sugar production (2)

Hydrochloric acid X Gelatin production

For regulating the pH value of the brine used in the production of cheese: Gouda, Edamer, Maasdammer, Boerenkaas, Friese and Leidse Nagelkaas

Barley extract X Only for antimicrobial use in sugar production

Pine nut extract X Only for antimicrobial use in sugar production.

If possible, from organic production

Ammonium hydroxide X Gelatin production

Hydrogen peroxide X Gelatin production

3. Permitted additives used for the production of yeast and yeast products in organic production

Name Basic yeast Yeast sweets/
preparation Special conditions
Calcium chloride X
Carbon dioxide X X
Citric acid X For pH regulation in yeast production
Lactic acid X For pH regulation in yeast production
Nitrogen X X
Oxygen X X
Potato starch X X For filtration
Only when obtained from organic production
Sodium carbonate X X For pH regulation
Vegetable oils X X Lubricant, glazing or anti-foaming agent
Only when obtained from organic production
Name Plant-based food preparation Preparation of food of animal origin" Special conditions
Carbon dioxide X X
Nitrogen X X
Ethanol X X Solvent
Tannic acid X Filtering agent
Egg white albumin X
Casein X
Gelatin X
Fish bladder gelatin
(isinglass) X
Vegetable oils X X Lubricants, glazing agents or anti-foaming agents
Only if derived from organic products
Silica gel
or colloidal solution X
Activated charcoal X
Talc X In accordance with the specific purity criteria for food additive E 553b
Bentonite X X Mead thickener (1)
Cellulose X X Gelatin production (1)
Diatomaceous earth X X Gelatin production (1)
Perlite X X Gelatin production (1)
Hazelnut shell X

Rice flour X Release agent
Organically grown beeswax
Beeswax X Release agent
Only if derived from organic products
Acetic acid/vinegar X Only if obtained from organic production
For fish processing and obtained by natural fermentation
Acetic acid obtained from or using GMOs is not
allowed
Thiamine hydrochloride X X Only for use in the processing of fruit wines, including cider, sparkling pear brandy and mead
Diammonium phosphate X X Only for use in the processing of fruit wines, including cider, sparkling pear brandy and mead
Wood fibres X X The wood raw material should come from a tree that has been sustainably harvested and whose logging has been approved.
The wood used must not contain toxic components. (treated after felling or containing natural toxins or toxins from microorganisms)
Notes:
(1) The restriction applies only to products of animal origin.
(2) The restriction applies only to products of plant origin.

ANNEX 10

Permitted ingredients of agricultural origin which are not produced by organic production methods

1. Unprocessed plant products and products derived therefrom

1.1. Edible fruit, stone fruit and seeds:

Acorn	Quercus spp.
Kola nut	Cola acuminata
Gooseberry	Ribes uva-crispa
Passion fruit	Passiflora edulis
Raspberries (dried)	Rubus idaeus

1.2. Edible spices and herbs:

Pepper (Peruvian)	Schinus molle L.
Horseradish seeds	Armoracia rusticana
Galangal	Alpinia officinarum
Saffron flowers	Carthamus tinctorius
Watercress	Nasturtium officinale

1.3. Other: Algae, including seaweed, permitted in conventional food preparation

2. Vegetable products

2.1. Vegetable fats and oils, whether or not refined or chemically modified, with the exception of the following

Cocoa	Theobroma cacao
Coconut	Cocos nucifera
Olive	Olea europaea
Sunflower	Elaeis guineensis
Palm tree	Elaeis guineensis
Rapeseed	Brassica napus, rapa
Saffron	Carthamus tinctorius

Sesame	Sesamum indicum
Soybean	Glycine max

2.2. The following sugars, starches and other products from cereals and potatoes:

Fructose
Rice husks
Unleavened bread
Rice and maise starch, not chemically modified

2.3. Other:

Pea proteins Pisum spp.
Rum, obtained only from sugar cane juice
Kirsch prepared with a base of fruit and flavourings

3. Animal products

Aquatic organisms other than those originating from aquaculture permitted in the preparation of food not obtained by organic production methods:

Gelatine
Whey powder "herasuola"
Intestines

ANNEX 11

Products and substances permitted for use in the organic production of products from the wine sector

Type of use according to the Regulations on oenological processes and oenological means for the production of must, wine and other products (Official Gazette of the Republic of Serbia, No. 26/15, 93/15, 41/17 and 84/18; hereinafter referred to as the Regulations) Product or substance name Special conditions and limitations of application within the framework of the conditions and limitations specified in the special regulation regulating the given area

Ventilation (aeration) or the use of gaseous oxygen (oxygenation),
According to Annex 2, point 1.1 of the Regulations– air
– gaseous oxygen -
Centrifugation and filtration
According to Annex 2, point 1.3 of the Regulations– perlite
– cellulose
– diatomaceous earth   Use only as an inert filtration medium

Use for the creation of an inert atmosphere and processing of products without the presence of air According to Annex 2, point 1.4 of the Regulations –carbon dioxide

– nitrogen
– argon
Use of yeast for the production of wines, dry or in suspension with wine;
According to Annex 2, point 1.5 of the Regulations
and
Use of preparations of yeast cell membranes;
According to Annex 2, point 1.14 of the Regulations
For dry wines, use of fresh, undiluted and healthy lees containing yeast from the recent vinification of dry wine.
According to Annex 2, point 1.20 of the Regulations – yeast * – yeast cell membranes

Use of one or more substances, with the possible addition of microcrystalline cellulose as an auxiliary substance, to stimulate the growth of yeasts, according to Annex 2, point 1.6 of the Regulations – diammonium phosphate – thiamine hydrochloride

– autolysed yeast cell walls

Use of certain substances

According to Annex 2, point 1.7 of the Regulations– sulfur dioxide

– potassium bisulfite or

potassium metabisulfite (1) The content of total sulfur dioxide in wine for direct human consumption must not exceed

– 100 mg/l for red wine if the sugar content is less than 2 g/l;

– 150 mg/l for white and rosé wine if the sugar content is less than 2 g/l.

(2) For all other wines, the maximum total sulphur dioxide content shall be reduced by 30 mg/l from that prescribed by a special regulation on the quality of such wines and permitted oenological practices

Charcoal treatment for oenological use

According to Annex 2, point 1.9 of the Regulations– coal for oenological use -

Type of use according to the Regulations on oenological processes and oenological means for the production of must, wine and other products (Official Gazette of the Republic of Serbia, No. 26/15, 93/15, 41/17 and 84/18; hereinafter referred to as the Regulations) Product or substance name Special conditions and limitations of application within the framework of the conditions and limitations specified in the special regulation regulating the given area

Clarification using one or more substances for oenological use

According to Annex 2, point 1.10 of the Regulations edible gelatin*

- vegetable proteins from
- wheat or peas*
- fish bladder*
- egg albumin*
- tannins*
- potato proteins*
- yeast protein extracts*
- casein and potassium
- caseinate
- chitosan isolated from Aspergillus niger,
- silicon dioxide
- bentonite
- pectolytic enzymes

Acidification

According to Annex 2, point 1.12 of the Regulations – lactic acid – L(+) tartaric acid

Deacidification

According to Annex 2, point 1.13 of the Regulations – L(+) tartaric acid

- calcium carbonate
- neutral potassium tartrate
- potassium bicarbonate

Use of lactic acid bacteria

According to Annex 2, point 1.16 of the Regulations – lactic acid bacteria

Addition of L-ascorbic acid

According to Annex 2, point 1.18 of the Regulations – L ascorbic acid

Foaming

According to Annex 2, point 1.21 of the Regulations – nitrogen

Addition of carbon dioxide

According to Annex 2, point 1.22 of the Regulations – carbon dioxide

Addition of citric acid for wine stabilisation

According to Annex 2, point 1.23 of the Regulations – citric acid

Addition of tannins

According to Annex 2, point 1.24 of the Regulations – tannins*

Addition of metavic acid

According to Annex 2, point 1.26 of the Regulations – metatartaric acid

Use of gum arabic

According to Annex 2, point 1.27 of the Regulations – gum arabic*

Use of substances to promote the precipitation of tartrates

According to Annex 2, point 1.29 of the Regulations – potassium bitartrate

Use of substances to eliminate deficiencies in terms of taste or smell of wine

According to Annex 2, point 1.30 of the Regulations – copper citrate

Addition of yeast mannoproteins to achieve stability of tartrate and protein in wine According to Annex 2, point 1.33 of the Regulations–yeast mannoproteins

Type of use according to the Regulations on oenological processes and oenological means for the production of must, wine and other products (Official Gazette of the Republic of Serbia, No. 26/15, 93/15, 41/17 and 84/18; hereinafter referred to as the Regulations) Product or substance name Special conditions and limitations of application within the framework of the conditions and limitations specified in the special regulation regulating the given area

Use of oak chips in the production and ageing of wines, including the fermentation of fresh grapes and musts, according to Annex 2, point 1.36 of the Regulations – oak shavings

("chips")

- Use of potassium alginate

According to Annex 2, point 1.37 of the Regulations – potassium alginate

- Processing using chitosan isolated from Aspergillus niger According to Annex 2, point 1.41 of the Regulations – chitosan isolated from Aspergillus niger

- Use of inactivated yeasts

According to Annex 2, point 1.45 of the Regulations – inactivated yeast

Note:

*If possible, obtained from organically produced raw materials.

ANNEX 12

Appearance of the national symbol for an organic product

National symbol of the Republic of Serbia for an organic product

ОРГАНСКИ ПРОИЗВОД

ПРОИЗВОД
ИЗ ПЕРИОДА
КОНВЕРЗИЈЕ

www.ingramcontent.com/pod-product-compliance
Lightning Source LLC
LaVergne TN
LVHW022103200926
848481LV00003B/1054
* 9 7 9 8 8 9 9 6 6 1 3 7 2 *